THE STRUCTURE OF IDEALIZATION

SYNTHESE LIBRARY

STUDIES IN EPISTEMOLOGY,

LOGIC, METHODOLOGY, AND PHILOSOPHY OF SCIENCE

VOLUME 139

THE STRUCTURE OF IDEALIZATION

Towards A Systematic Interpretation Of The Marxian Idea Of Science

by

LESZEK NOWAK

Springer-Science+Business Media, B.V

Library of Congress Cataloging in Publication Data

Nowak, Leszek.
The structure of idealization.

(Synthese library; v. 139)
Bibliography: p.
Includes index.
1. Science—Philosophy. 2. Marx, Karl, 1818–1883.
I. Title.
Q175.N67 501 79–20656

ISBN 978-90-481-8356-2 ISBN 978-94-015-7651-2 (eBook)
DOI 10.1007/978-94-015-7651-2

Softcover reprint of the hardcover 1st edition 1980

TABLE OF CONTENTS

INTRODUCTION

Much is said in Marxist literature about Marxist methodology which is supposed to be entirely original – differing a great deal from all other trends in the modern philosophy of science. On the other hand, however, it is unfallacious to state that there are no people outside Marxism who would like to deny this statement. This has to put those who really believe that Marxism has something important to say in philosophy of science on guard: if someone says something important others usually are inclined to protest. But who is inclined to protest when it is stated that Marx employed both induction and deduction, a historical method and a logical one as well, synthesis, but also analysis, etc? Who is inclined to protest when it is not known what within this framework 'induction', 'deduction' 'history' or 'logic' mean? Who is inclined to protest when 'Marxist methodology' is presented not with the aid of precise definitions and clear hypotheses but with the aid of a jungle of quotations?

I think that the main malfeasance of the current 'Marxist methodology', is that of ecclecticism. The methodology of Marx is presented as a collection of trivial and/or obscure ideas but not as a system of statements subordinated to any clear, definite viewpoint presenting a new grasp of the nature of scientific cognition. Search for reconstruction of Marxian methodology as a system of the kind is the main aim of this book.

Being conscious of these preliminary points, it becomes clear that there is no infallible way of finding out what 'Marxist methodology' states. The only way of finding out whether there is a Marxist methodology or not is to propose interpretational hypotheses concerning Marx's own scientific practice, to test them in the light of the analysis of his works, and to compare them with what had been explicitly stated by the author of *Capital*.

This constitutes the first step (see Part I of the book). The problem arises as to whether the methodological principles employed in the classic works of Marxism are really original in comparison with modern trends of philosophy of science. If it is demonstrated that the Marxian methodological principles, confirmed by his remarks about scientific method, differ significantly from those provided by other methodological schools, then it will be proven that there is in fact a Marxist methodology. This will constitute the second step (see Part II of the book).

The third step is not interpretational. It consists in an attempt at doing what the classics did not do at all, an attempt at the construction of systematic theory of science. It is obvious that different theories even based on the same reconstruction of Marxian ideas may be constructed in a number of ways. Therefore what is presented in Part III of the book is just one of possible interpretations of the Marxist methodology. It is *a* Marxist methodology since it aims at developing of what Marx had been saying. It is not *the* Marxist methodology because there are many different admissible competing interpretations of Marx's ideas and there can be many different developments of an interpretation of the classics' writings. The way of deciding which of those possible Marxist philosophies of science is the best one is only one: by explaining as much from the traditional ideas of Marxism as possible. My arguments supporting *my* reconstruction, are contained in Part IV of the book.

ACKNOWLEDGEMENTS

Part of the material in this volume has been published elsewhere, as follows:

Chapter 2, Section 5, and Chapters 16, 17, and 18: *Poznań Studies in the Philosophy of the Sciences and the Humanities*, Vol. 1, No. 3, Grüner, Amsterdam 1975, pp. 39–46, 49–59.

Chapter 3, Sections 1, 2 and 4: *Marx' Methodologie, Neue Hefte für Philosophie*, Heft 13, Vandenhoeck and Ruprecht, Göttingen, 1978, pp. 81–91.

Chapter 9, Section 2: *Philosophy of Science*, Vol. 39, No. 4, December 1972, pp. 538–546.

Chapter 10, Section 2: *Ibid.*, pp. 546–548.

Chapter 12, Sections 1 and 2: *Poznań Studies in the Philosophy of the Sciences and the Humanities*, Vol. 2, No. 3, Grüner, Amsterdam, 1976, pp. 81–95.

Chapter 13, Section 2, and Chapter 15: *Ibid.*, pp. 17–26.

Chapter 13, Section 3: *Philosophy of Science*, Vol. 42, No. 2, June, 1975, pp. 192–201.

Chapter 14, Sections 3 and 4: *Poznań Studies in the Philosophy of the Sciences and the Humanities*, Vol. 2, No. 1, Grüner, Amsterdam, 1976, pp. 54–62.

I am grateful to all the editors and publishers concerned for their permission to use these materials.

Many colleagues have helped me by discussing the topics and reading the manuscript. I am very grateful to Jerzy Brzeziński, Jolanta Burbelka, Piotr Buczkowski, Andrzej Klawiter, Krzysztof Łastowski, Izabella Nowakowa, Wojciech Patryas, Mirosław Stefański, Barbara Tuchańska and Renata Zielińska for improving the contents of the book. And to Catherine Modelski-Gussow for improving the English. Needless to say, the full responsibility for the book is mine.

PART I

THE METHODOLOGY OF *CAPITAL*. AN ATTEMPT AT A RECONSTRUCTION

CHAPTER 1

CASE STUDY: THE THEORY OF VALUE

1. THE LAW OF VALUE

Let us consider a typical theorem of Marx in *Capital*, the law of value. It says that the prices of commodities are formed with respect to their values (the amount of socially indispensable time to produce these commodities). But that holds only under certain conditions. Let us try to reconstruct them.

This is the simplified formulation of the law of value from Marx's lecture for workers:

(1) "if supply and demand equilibrate each other, the market prices of commodities will correspond (. . .) with their values" (Marx (4), pp. 53–54).

In other words, the formula (1) states that:

(2) if x is a commodity and the difference between the demand for x and the supply of x equals zero, then the price of the commodity x depends in a linear way on the value of that commodity.

Let us formulate the statement (2) in a shorter form:

(3) if $C(x)$ and $D(x) - S(x) = 0$, then $P(x) = l(V(x))$,

where $C(x)$ means that x is a commodity, D stands for the demand for x, S stands for its supply, P is the price of x and V its value; l denotes a linear function.

Now one may see that statement (3) is based on the assumption of rather strange kind:

(4) $D(x) - S(x) = 0.$

This does not hold in reality, because changes of demand and supply are an ordinary phenomenon in a free competition economy. In other words, there is no economic system which would have the demand and supply of any commodity balanced all the time. Therefore, the statement (3) does not describe the phenomenon of creating prices, since the latter, as Marx perfectly knew, occur in a situation of a permanent imbalance between supply and demand. And Marx does not only formulate the statement (1) (and, tacitly, (3)) but even calls it the basic law for the capitalist economy! The explanation can be only one: the statement (1) is not, according to him, a description of the course of economic events. But why then does he call it a "law"?

To look for an answer to this question let us try to reconstruct the law of value in a systematic way. Let us note, first of all, that the counterfactual nature of the condition (4) is not a matter of accident. Here are Marx's comments as to the status of the law of value:

Here we need only consider the forms which capital passes through in the various stages of its development. The *real conditions* within which the actual process of production takes place *are* therefore *not analysed*. It is assumed throughout, that *the commodity is sold at its value*. We do not examine the competition of capitals, nor the credit system, nor the actual composition of society, which by no means consists only of two classes, workers and industrial capitalists (Marx (5), Vol. II, pp. 492–493.) [Italics – L.N.]

Elsewhere, Marx gives the following explanation:

If prices actually differ from values, we must, first of all, reduce the former to the latter, in other words, treat the difference as accidental in order that the phenomena may be observed in their purity, and our observations not interfered with by disturbing circumstances that have nothing to do with the process in question. (Marx (1), Vol. I, p. 166.)

One may infer two conclusions from what has been said: firstly, the counterfactual nature of the assumption (4) of the law of value is deliberate on the part of Marx, and, secondly, the assumption is not the only condition of the kind.

The first condition of the law of value to be specified here will be an assumption which in fact is a combination of several conditions. According to it, the economic system analysed by Marx in *Capital* is that of perfect competition. First of all, this means that the producers of commodities act rationally from the economic point of view, that their primary goal can be reduced to the production of a maximum of surplus value for a given amount of capital invested. In Marx's words:

It is the constant aim of capitalist production to produce a maximum of surplus-value or surplus-product with the minimum[1] capital outlay; and to the extent that this result is not achieved by overworking the workers, it is a tendency of capital to seek to produce a given product with the least possible expenditure–economy of power and expense. It is therefore the economic tendency of capital which teaches humanity to husband its strength and to achieve its productive aim with the least possible expenditure of means. (Marx (5), Part II, pp. 547–548.)

The second condition of this assumption presupposes that there is no monopoly in economy (Marx (1), Vol. III, p. 192):

apart from the effect of monopolies and some other modifications I must now pass by, all descriptions of commodities are, on the average, sold at their respective *values* (Marx (4), p. 54.)[2]

Finally, the assumption in question presupposes the tacit condition that the means of production as well as the manpower which serves them are perfectly mobile, so the producers do not face any difficulties either in shifting production to turning out other kinds of commodities or in transferring their factories to other places (Marx (1), Vol. III, pp. 192–193).

The second assumption to be specified here is the following:

We here take no account of export trade (. . .) . In order to examine the object of our investigation in its integrity, free from all disturbing subsidiary circumstances, we must treat the whole world as one nation, and assume that capitalist production is everywhere established and has possessed itself of every branch of industry. (Marx (1), Vol. I, p. 581.)

According to the third assumption,

in all those spheres of production into which social labour in a given country falls we have (. . .) the same rate of surplus value. (*Ibid.*, Vol. III, p. 131.)

Let us add that the rate of surplus value is the quotient of surplus value by variable capital (the value of manpower).

Marx further assumes that

landed property is taken as = 0, that is, landed property as a particular economic relation does not yet concern us. (Marx (2), pp. 97–98).

Here is justification for the assumption:

In all social forms a given type of production determines the rank and importance of all other types, and hence relations in that type of production determine the rank and importance of all other relations (. . .). In bourgeois society (. . .) agriculture is on increasing scale becoming just a branch of industry and is totally dominated by capital. The same applies to ground rent (. . .). *Ground rent cannot be understood without capital. But capital can be understood without ground rent.* Capital is that economic force in bourgeois society which dominates everything. It must be both the point of departure and the point of arrival and must be analysed before landed property is analysed. (Marx (3), p. 330) [italics – L.N.]

This is the reason why Marx eliminates the economic peculiarities of agriculture as a branch of production. These special characteristics of agriculture reduce to the fact that the organic composition of capital (the quotient of constant capital, i.e., the value of means of production by variable capital) in agriculture is lower than the composition of socially average capital (Marx (1), Vol. III, p. 741). Hence, in order to eliminate economic peculiarities of agriculture Marx assumes that the organic composition of agricultural capital is the same everywhere.

Marx, in turn, adopts the assumption that commercial profit on the commodities produced in the economic system under consideration equals zero (*ibid.*, p. 265). This means that the functions which in real economic life are performed by merchants as a specialized group of capitalists, are, under the assumption in question, performed by the producers themselves.

The above two assumptions concerning ground rent and commercial profit are only some special cases of a larger group of assumptions adopted in the first volume of *Capital*:

> The capitalist who produces surplus-value – i.e., who extracts unpaid labour directly from the labourers, and fixes it in commodities, is, indeed, the first appropriator, but by no means the ultimate owner, of this surplus value. He has to share it with capitalists, with landowners, etc., who fulfil other functions in the complex of social production. Surplus-value, therefore, splits up into various parts. Its fragments fall to various categories of persons, and take various forms, independent the one of the other, such as profit, interest, merchants' profit, rent, etc. *It is only in Book III, that we can take in hand these modified forms of surplus-value.* (. . .) [Hence] *we treat the capitalist producer as owner of the entire surplus-value,* or, better perhaps, as the representative of all sharers with him in the booty. We, therefore, first of all consider accumulation from an abstract point of view–i.e., as a mere phase in the actual process of production (Marx (1), Vol. I, pp. 564–565).

The reason for doing so is that

> *the simple fundamental form of the process of accumulation is obscured* by the incident of the circulation which brings it about, and *by the splitting up of surplus-value.* An exact analysis of the process, therefore, demands that we should, for a time, disregard all phenomena that hide the play of inner mechanism. (*Ibid.*, p. 565.) [All italics – L.N.]

This is an important point, because "All economists share the error of examining surplus-value not as such, in its pure form, but in the particular forms of profit and rents" (Marx (5), Part I, p. 40.)

Another of Marx's assumption is that the average organic compositions of capitals in all branches of production of the economic system under consideration are the same (Marx (1), Vol. III, p. 60). In other words, Marx presupposes that each branch of production in a given economic system characterizes itself by the same average organic composition of capital.

Marx, in turn, adopts the condition that demand for each commodity equals its supply (*ibid.*, p. 186). The reason for assuming this is the following:

Supply and demand regulate nothing but the temporary *fluctuations* of market prices. They will explain to you why the market price of a commodity rises above or sinks below its *value*, but they can never account for that *value* itself. Suppose supply and demand to equilibrate, or, as the economists call it, to cover each other. The very moment these opposite forces become equal they paralyse each other, and cease to work in the one or the other direction. At the moment when supply and demand equilibrate each other, and therefore cease to act, the *market price* of a commodity coincides with its *real value*, with the standard price round which its market prices oscillate. In inquiring into the nature of that *value*, we have, therefore, nothing at all to do with the temporary effects on market prices of supply and demand. (Marx (4), p. 45.)

And even more clearly:

In reality, supply and demand never coincide, or if, they do, it is by a mere accident, hence scientifically = 0, and to be regarded as not having occurred. But political economy assumes that supply and demand coincide with each other. Why? To be able to study phenomena in their fundamental relations, in the form corresponding to their conception, that is, to tstudy hem independent of the appearances caused by the movement of supply and demand. (Marx (1), Vol. III, p. 186.)

And finally, the last assumption to be taken here into account is that the organic compositions of the capitals engaged in any given branch of production are the same (Marx (1), Vol. III, p. 38).

It must be added that the assumptions specified above are not the only ones adopted by Marx in connection with the law of value. He also assumes that the only type of money is gold ("Throughout this work, I assume, for the sake of simplicity, gold as the money-commodity" – *ibid.*, Vol. I, p. 94), a constant value of money ("we shall consider the value of gold to be given" – *ibid.*, p. 118), the same conditions of the circulation of capital, non-existence of interest on the capital taken into consideration (and hence non-existence of bankers as a specialized group of capitalists), etc.

The eight assumptions specified above must, at any rate, be formulated more precisely. Since the law of value states that the price of a commodity x equals the value of the commodity x, hence all those assumptions must be formulated so that they be restricted to that variable.

$$p_1: PC(x),$$

means that the commodity x is produced in the conditions of perfect competition.

$$p_2: E(x) - I(x) = 0,$$

means that the difference between the exports and the imports of a given economic system producing the commodity x equals zero.

$$p_3: m'(x) - M'(x) = 0$$

i.e., the rate of surplus value characteristic of the capital used to produce the commodity x is identical with the average rate of surplus value characteristic of the capital engaged in the economic system in question as a whole (since the value of the difference of these two magnitudes equals zero).

$$p_4: ORG_A(x) - ORG_I(x) = 0,$$

i.e., the difference between the average organic composition of agricultural capitals and the average organic composition of remaining capitals in a given economic system producing the commodity x equals zero.

$$p_5: mer(x) = 0,$$

that is, the merchant profit on the commodity x equals zero.

$$p_6: ORG(x) - ORG_E(x) = 0,$$

i.e., the difference between the average organic composition of capital in that branch of production which turns out the commodity x and the average organic composition of capital in the whole economy in question equals zero.

$$p_7: D(x) - S(x) = 0,$$

i.e., the difference between the supply of the commodity x and the demand for that commodity equals zero.

$$p_8: Org(x) - ORG(x) = 0,$$

i.e., the difference between the organic composition of the capital used in the production of the commodity x and the average organic composition of capital in that branch of production which turns out the commodity x is equal to zero.

Let us add another condition of a quite different nature than $p_1 - p_8$

which has been tacitly assumed above. The condition establishes the universe of discourse telling that the object x is a commodity:

$$p_0\colon C(x).$$

Now the law of value can be formulated as follows:

(T^8) if $C(x)$ and $PC(x)$ and $E(x) - I(x) = 0$ and $m'(x) - M'(x) = 0$ and $ORG_A(x) - ORG_I(x) = 0$ and $mer(x) = 0$ and $ORG(x) - ORG_E(x) = 0$ and $D(x) - S(x) = 0$ and $Org(x) - ORG(x) = 0$,
then $P(x) = l(V(x))$.

This formulation is a much better approximation to Marxian law of value contained in *Capital* than the formulation (3) we have started with. But the methodological nature of the statement T^8 is the same: it presupposes conditions $p_1 - p_8$ that are not satisfied by any real commodity. There is no economic system which would be in perfect competition, which would be closed, which would have agriculture equalled with industry as to the organic composition of capital, etc. Such an economic system resembles ideal gases, perfectly rigid bodies, and other constructs of the type. In other words, the law of value is satisfied vacuously in the empirical domain investigated by Marx. It is therefore a counterfactual statement.

But, what is more, this counterfactual nature of the basic law of *Capital* was intended by Marx. He built that law according to his own principle of abstraction. As he writes,

> The physicist either observes physical phenomena where they occur in their most typical form and most free from disturbing influence, or, wherever possible, he makes experiments under conditions that assure the occurrence of the phenomenon in its normality. (Marx (1), Vol. I, p. 8.)

But

> In the analysis of economic forms (. . .) neither microscopes nor chemical reagents are of use. The force of abstraction must replace both. (*Ibid.*, p. 8.)

"The force of abstraction" is indispensable for building economic laws:

> One can see that though Ricardo is accused of being too abstract, one would be justified in accusing him of the opposite: lack of the power of abstraction, inability, when dealing with the values of commodities, to forget profits, a factor which confronts him as a result of competition (Marx (5), Part II, p. 191.)

The vulgar mob has (. . .) concluded that theoretical truths are abstractions which are at variance with reality, instead of seeing, on the contrary, that Ricardo does not carry true abstract thinking far enough and is therefore driven into false abstraction. (*Ibid.*, p. 437.)

On the other hand, "abstract laws" are necessary as the point of departure to build less abstract (and more realistic) laws that are better approximations to real economic phenomena. This is Marx's own comment about the role of the law of value:

Science consists precisely in demonstrating *how* the law of value manifests[3] itself. So that if one wanted at the very beginning to 'explain' all the phenomena which seemingly contradict that law, one would have to present the science *before* science. It is precisely Ricardo's mistake that in his first chapter on value he takes *as given* a variety of categories that have not yet been explained in order to prove their conformity with the law of value. (Marx (6), p. 196.)

This is, according to Marx, a usual scientific procedure:

To explain (. . .) the *general nature of profits*, you must start from the theorem that (. . .) commodities are *sold at their real value*, and that *profits are derived from selling them at their values*, that is, in proportion to the quantity of labour realised in them. If you cannot explain profit upon this supposition, you cannot explain it at all. This seems paradoxical and contrary to everyday observation. It is also paradoxical that the earth moves round the sun, and that water consists of two highly inflammable gases. Scientific truth is always paradoxical, if judged by everyday experience, which catches only the delusive appearance of things (Marx (4), p. 54.)

Now, we shall try to reconstruct Marx's way of "demonstrating how the law of value manifests itself" and to reveal the hidden structure it contains.

2. TRANSFORMATIONS OF THE LAW OF VALUE IN *CAPITAL*

The law of value T^8 is a very abstract economic statement resembling, as we shall see later, abstract laws of physics, the laws of perfect gases, the laws of perfectly rigid bodies, etc. Now, the law of value T^8 is preserved in this abstract form up to Chapter X in Vol. III of Marx's *Capital*. In that chapter, he removes the assumption p_8, that is, he admits the possibility that there are differences in the organic compositions of capital engaged in various branches of production. In other words, he replaces the condition p_8 with the condition:

$$\text{non-}p_8\text{: } Org(x) - ORG(x) \neq 0,$$

i.e., the difference between the organic composition of capital used in the production of the commodity x and the average organic composition of capital in that branch of production results in the commodity x differing from zero. Hence, the organic composition of the capital producing x and the average organic composition in a given branch of production may differ from what is in reality. In this respect, market conditions described in Chapter X in Vol. III of *Capital* resemble empirical market conditions.

Removing the condition p_8, Marx introduces distinction between the individual value and the market value of a commodity:

There is also the *market value* (. . .) to be distinguished from the individual value of particular commodities produced by different producers (. . .) market value is to be viewed as the average of commodities produced by different producers (Marx (1), Vol. III, p. 175.)

By stating that "the market value (...) forms of centre of fluctuations for market prices" (*ibid.*), Marx formulates the transformation of the law T^8 being in force under conditions $p_1 - p_7$ and non $- p_8$. The form of that transformation depends on the relation between the organic composition of the capital producing a given commodity and the average organic composition in a given branch of production. In connection with this Marx says:

The individual value of some of these commodities will be below their market value (that is, less labour time is required for their production than expressed in the market value) while that of others will exceed the market value. (*Ibid.*).

Finally, some commodities have their individual value equal to the market value in a given sphere of production: this applies to "the commodities produced under average conditions of their respective sphere" (*ibid.*). Since "the conditions of production" are nothing else than a synonym for "the organic composition of capital", hence the same may be formulated thus: when the organic composition of the capital engaged in the production of the commodity x is equal to, higher than, or lower than, the organic composition of the average capital engaged in a given sphere of production of the said commodity, then the individual value of the commodities produced by that capital is, accordingly, equal to, or lower than, or higher than, the market value of the commodity x. Hence it may be said that Marx assumes the following transformation of the law of value:

(T^7) if $C(x)$ and $PC(x)$ and $E(x) - I(x) = 0$ and $m'(x) - M'(x) = 0$ and $ORG_A(x) - ORG_I(x) = 0$ and $mer(x) = 0$ and $ORG(x) - ORG_E(x) = 0$ and $D(x) - S(x) = 0$

and $Org(x) - ORG(x) \neq 0$,
then $P(x) = l(V(x)) + a(Org(x) - ORG(x))$

where the function a satisfies the following conditions:

(1) $a(Org(x) - ORG(x)) = 0$, if $Org(x) = ORG(x)$,
(2) $a(Org(x) - ORG(x)) < 0$, if $Org(x) > ORG(x)$,
(3) $a(Org(x) - ORG(x)) > 0$, if $Org(x) < ORG(x)$.

The statement T^7 may be termed the law of market value. It states that in less abstract conditions, namely those characterized by the assumptions $p_1 - p_7$, the prices of commodities are not simply equal to the values of those commodities, but are equal to their market values, that is, values modified according to the organic composition of the capital engaged in the production of those commodities. It must be emphasized that the function a is only partly defined that is, it is defined as to its type only, because conditions (1) – (3) are satisfied by the whole class of functions. This, however, is typical of the laws formulated in the "qualitative" language of economics: an entire class of mathematical functions proves adequate for their formulation (Czerwinski (1), p. 143; Allen (1), pp. 112, 114).

Now, the statement T^7 is a less abstract and, at the same time, a more realistic economic theorem. It contains less conditions of the abstract type in its antecedent and the correction connected with the removed assumption p_8 is introduced in its consequent. The statement T^7 is a more realistic and a more complicated economic theorem than the law of value T^8.

Subsequently, Marx removes the assumption p_7, which postulates that the supply of every commodity equals the demand for that commodity. He thus comes to analyse still less abstract conditions, namely those characterized by the assumptions $p_1 - p_6$. In that economic system, the commodities satisfy the assumptions $p_1 - p_6$, the assumption non–p_8 and additionally the assumption:

$$\text{non} - p_7\colon D(x) - S(x) \neq 0.$$

Hence, in these two respects the commodities now under consideration, bear an analogy to real commodities. A corresponding correction is also introduced by Marx concerning the law of market value T^7 and resulting from the removal of the condition p_7 stating that demand and supply are equal to each other. As in the previous case, that correction is defined as to its type. Marx says:

> For the market price of identical commodities, each, however, produced under different individual circumstances, to correspond to the market value and not to deviate from it either by rising above or falling below it, it is necessary that the pressure exerted by different sellers upon one another be sufficient to bring enough commodities to market to fill the social requirements (. . .). Should the mass of products exceed this demand, the commodities would have to be sold below their market value; and conversely, above the market value if the mass of products were not large enough to meet the demand. (Marx (1), Vol. III, p. 177.)

Thus "supply and demand regulate the market price, deviations of the market price from the market value" (*ibid.*, p. 178.). Marx, then, assumes that the next transformation of the law of value is dependent on the relationship between demand and supply:

(T^6) if $C(x)$ and $PC(x)$ and $E(x) - I(x) = 0$ and $m'(x) - M'(x) = 0$ and $ORG_A(x) - ORG_I(x) = 0$ and $mer(x) = 0$ and $ORG(x) - ORG_E(x) = 0$, and $D(x) - S(x) \neq 0$ and $Org(x) - ORG(x) \neq 0$,
then $P(x) = l(V(x)) + a(Org(x) - ORG(x)) + b(D(x) - S(x))$,

where the function b satisfies the following conditions:

(1) $b(D(x) - S(x)) = 0$, if $D(x) = S(x)$,
(2) $b(D(x) - S(x)) < 0$, if $D(x) < S(x)$,
(3) $b(D(x) - S(x)) > 0$, if $D(x) > S(x)$.

This theorem may be termed the law of market value under the consideration of demand and supply. It states that in more realistic conditions, namely those characterized by the assumptions p_1 p_6, the prices of commodities are not equal to their market values, but are equal to their market values modified with respect to fluctuations of demand and supplay; the latter factor, which is a very characteristic property of each market, is taken into account in the Theorem T^6. The Theorems T^8 and T^7 are concerned with ideal market where fluctuations of demand and supply do not occur, but the theorem T^6 describes such a market which is idealized with respects $p_1 - p_6$, but not with respect to p_7; therefore it takes into account the fluctuations of demand and supply.

Note that the assumption p_6 which postulates the equality of average organic composition of capital engaged in various sphere of production, together with the assumption p_4 (which postulates the equality of average organic composition of capitals engaged in agriculture and outside agriculture) and the assumption p_3 (which postulates the equality of the rate of surplus value in the whole of a given national economy), result in the conclusion that in the economic system under consideration there is no competition between various branches of the economy, i.e., there is no flow of capital from one branch to another. This is so because it follows from Marx's formula for the rate of profit p':

$$p' = m'/((c/v) + 1)$$

(where c/v stands for the ratio of constant capital c to variable capital v, i.e., the organic composition of capital, symbolized above as *Org* or *ORG*), together with the assumptions listed above, that the rate of profit is the same in all spheres of production. And if that is so then, under the assumption p_1, capitalists do not transfer their capitals from one branch of production to another as this does not increase their profits.

A change occurs when Marx removes the assumption p_6 and accordingly takes into consideration more realistic conditions which are described by the assumptions $p_1 - p_5$ and non-p_7 and non–p_8, and also:

$$\text{non–}p_6\text{: } ORG(x) - ORG_E(x) \neq 0,$$

i.e., the difference between the average organic composition of capital in that branch of production which turns out the commodity x and the average organic composition of capital in the whole economy in question differs from zero. Then, under the assumptions accepted, among others the assumptions made about rational behaviour of capitalists and a lack of obstacles that would prevent the transfer of means of production and manpower from one sphere of production to another, we have to deal with competition between the various branches of production: capitalists transfer their capitals from spheres of production marked by a lower rate of profit to those marked by a higher rate of profit. Accordingly, there is an increased supply in the latter spheres, but since it is not balanced with a corresponding rise in demand, then the capitalists, in order to avoid over-production, must lower prices. In this way the rate of profit decreases in those branches which previously were marked by a high rate of profit. The reverse process takes place in those branches of production which were previously marked by a low rate of profit: an outflow of capitals results in decreased supply, and the capitalists raise prices, so that the rate of profit in those branches increases. Thus, in Marx's opinion, these two processes, directed contrary to one another, result in the equalization of rates of profit in the various spheres of production and in the formation of an average rate of profit for the national economy as a whole (or, as Marx also writes, a general rate of profit). According to Marx, that average rate of profit can be preserved only if prices of commodities equal a "transformed form of market value" which are prices of production.

What competition within *the same* sphere of production brings about, is the determination of the *value of the commodity in a given sphere* by the average labour-time required in it, i.e., the creation of the *market-value*. What competition between the *different* spheres of production brings about is the *creation of the same general rate of profit* in the *different* spheres through the levelling out of the different market-values into market-prices, which are *cost-prices* [prices of production – L.N.] that are different from the actual market-values. Competition in this second instance by no means tends to assimilate the prices of the commodities to their values, but on the contrary, to reduce their values to cost-prices [prices of production – L.N.] that differ from these values, to abolish the differences between their values and cost-prices [prices of production – L.N.]. (Marx (5), Part II, p. 208.)

When the assumption p_6 is removed, the competition between different spheres of production is taken into consideration, then prices of commodities do not correspond to the market-values of these commodities but to their prices of production.

What has been said here of market value applies to the price of production as soon as it takes the place of market value (. . .). And this price of production is, in its turn, the centre around which the daily market prices fluctuate. (Marx (1), Vol. III, p. 176.)

There are now three groups of branches of production, namely those with an average, a low, and a high organic composition of capital. In this connection,

The relation of the price of production of a commodity to its value is determined solely (. . .) by the organic composition of the capital producing it. If the composition of capital in a given sphere of production is lower than that of the average social capital (. . .), then the value of its product must lie above the price of production (. . .). The opposite is the case when the capital invested in a certain sphere of production is of a higher composition than the average social capital. The value of commodities produced by it lies below their price of production. (*Ibid.*, pp. 140–141.)

Marx refers here to market value, or even just to the value, but there is no doubt that he means what we have termed market value under the consideration of demand and supply. This follows from the presupposition of the consistency of his reasoning: as we have seen above, the concept of price of production assumes the working of demand and supply (since the mechanism of demand and supply shapes that average rate of profit which is to be guaranteed by the sale of commodities at prices equal to the prices of production), and hence it is a transformed form" (*ibid.*, p. 26) of market value, which takes demand and supply into consideration.

It must be realized that for Marx the value of a commodity is the sum of variable capital (v), constant capital (c), and surplus value (m), i.e., the sum of the cost of production:

$$k = v + c$$

and surplus value m:

$$V = k + m.$$

Hence the statement T^6 for instance can be rewritten in this way:

(T^6) if $C(x)$ and $PC(x)$ and $E(x) - I(x) = 0$ and $m'(x) - M'(x) = 0$ and $ORG_A(x) - ORG_I(x) = 0$ and $mer(x) = 0$ and $ORG(x) - ORG_E(x) = 0$ and $D(x) - S(x) \neq 0$ and $Org(x) - ORG(x) \neq 0$,
then $P(x) = l(k(x) + m(x)) + a(Org(x) - ORG(x)) + b(D(x) - S(x))$.

After the removal of the assumption p_6 Marx states that "The formula that the price of production of a commodity $= k + p$ (. . .) is now more precisely defined with $p = kp'$ (p' being the general rate of profit)" (Marx (1), Vol. III, p. 163). This correction depends on the difference between the average organic composition of the branch of production which produces a given commodity and the organic composition of capital which is average for the entire national economy. Therefore, the transformed form of the statement T^6 is:

(T^5) if $C(x)$ and $PC(x)$ and $E(x) - I(x) = 0$ and $m'(x) - M'(x) = 0$ and $ORG_A(x) - ORG_I(x) = 0$ and $mer(x) = 0$ and $ORG(x) - ORG_E(x) \neq 0$ and $D(x) - S(x) \neq 0$ and $Org(x) - ORG(x) \neq 0$,
then $P(x) = k(x) + k(x)p'(x)$.

The statement T^5 is a reconstruction of the law of production prices.

Marx in turn removes the assumption p_5, thus passing to a more realistic economic system, namely one which is characterized by the assumptions p_1, p_2, p_3, p_4, non-p_6, non-p_7, non-p_8 and additionally by the assumption:

non-p_5: $mer(x) > 0$,

i.e., the merchant rate of profit is greater than zero. The sociological, so to speak, meaning of the assumption is that the existence of commercial capital is admitted.

Commercial capital is, therefore, nothing, but the producer's commodity capital which has to undergo the process of conversion into money – to perform its function of commodity capital on the market – the only difference being that instead of representing an incidental function of the producer [on the assumption that commercial capital did not exist – L.N.], it is now the exclusive operation of a special kind of capitalist, the merchant. (Marx (1), Vol. III, p. 265.)

Now that merchant

makes his profit only during and through the circulation process, and only from an excess of his selling price over his purchase price. Yet he does not sell the commodities above their value, or above their price of production, precisely because he has bought them from the industrial capitalist below their value, or below their price of production (*ibid.*, p. 281).

This – Marx continues – leads to a closer and more comprehensive definition of the price of production. By price of production we mean, just as before [i.e., before the assumption which eliminated merchant profit had been removed – L.N.], the price of a commodity = its costs (. . .) + the average profit. But this average profit is now determined differently. It is determined by the total profit produced by the total productive capital; but not as calculated on the total productive capital alone [but] as calculated on the total productive + merchant's capital (. . .). The share of the total profit falling to merchant's capital is thus included in the average rate of profit. The actual value, or price of production, of the total commodity capital is therefore $k + p + m$ (where m is commercial profit) (. . .). We shall henceforth use the term price of production in this, more precise, sense. It is evident, therefore, that (. . .) commercial profit (. . .) equals the excess of the selling price over the price of production of the commodity which, for the merchant, is its purchase price; but that the actual price of the commodity = its price of production + the commercial profit. (*Ibid.*, pp. 280–281.)

As can be seen from the above, Marx assumes here the following modification of the statement T^5:

(T^4) if $C(x)$ and $PC(x)$ and $E(x) - I(x) = 0$ and $m'(x) - M'(x) = 0$ and $ORG_A(x) - ORG_I(x) = 0$ and $mer(x) \neq 0$ and $ORG(x) - ORG_E(x) \neq 0$ and $D(x) - S(x) \neq 0$ and $Org(x) - ORG(x) \neq 0$,
then $P(x) = k(x) + k(x)\,p'(x) + mer(x)$.

This statement will be called the law of price production under the consideration of merchant profit.

Marx now removes the assumption p_4, thus taking into account an essential economic peculiarity of agriculture, namely the fact that the organic composition of capital engaged in agriculture is much lower than the average organic composition of capital engaged outside agriculture.

then, *prima facie*, this expresses the fact that in countries with developed production agriculture has not progressed to the same extent as the processing industries (. . .). Whether the composition of agricultural capital is lower than that of the average social capital in a specific country where capitalist production prevails, for instance, England is a question which can only be decided statistically (. . .). In any case, it is theoretically established that the value of agricultural products can be higher than their price of production only on this assumption. (Marx (1), Vol. III, pp. 741–742.)

That "surplus value" is called absolute rent. The relationship between absolute rent and the ratio of the organic composition of agricultural capital to capital outside agriculture is stated by Marx as follows:

If the average composition of agricultural capital were equal to, or higher than, that of the average social capital, then absolute rent (. . .) would disappear. (*Ibid.*, p. 746.)

Marx explains in another place:

It is quite simply the *private ownership* of land, mines, water, etc. by certain people, which enables them to snatch, intercept and seize *the excess surplus-value over and above profit* (average profit, the rate of profit determined by the general rate of profit) contained in the commodities of these particular spheres of production, these particular fields of capital investment, and so to prevent it from entering into general process by which the general rate of profit is formed. Moreover, some of this surplus-value is actually collected in every industrial enterprise, since rent for the land used (by factory building, workhouses, etc.) figures in every instance, for even where the land is available free, no factories are built, except in the more or less populated areas with good means of communication (Marx (5), Part II, p. 37).

It may accordingly be assumed that Marx adopts the following transformation of the law of value:

(T^3) if $C(x)$ and $PC(x)$ and $E(x) - I(x) = 0$ and $m'(x) - M'(x) = 0$ and $ORG_A(x) - ORG_I(x) \neq 0$ and $mer(x) \neq 0$ $ORG(x) - ORG_E(x) \neq 0$ and $D(x) - S(x) \neq 0$ and $Org(x) - ORG(x) \neq 0$,
then $P(x) = k(x) + k(x)p'(x) + mer(x) + R(x)$,

where R stands for the absolute rent of a given commodity. In addition to that R depends on the difference between ORG_A and ORG_I:

$$R(x) = d(ORG_A(x) - ORG_I(x)),$$

where the function d satisfies in any case the following conditions:

(1) $d(ORG_A(x) - ORG_I(x)) = 0$, if $ORG_A(x) \geqslant ORG_I(x)$
(2) $d(ORG_A(x) - ORG_I(x)) > 0$, if $ORG_A(x) < ORG_I(x)$.

The statement (T^3) will be termed the law of prices of production under the consideration of merchant profit and rent.

At last Marx removes the condition p_3 adopting realistically that:

$$\text{non} - p_3: m'(x) - M'(x) \neq 0$$

I shall not discuss the latter point in detail (see Nowak (1), p. 124ff) but simply state that the removal of the condition p_3 leads to the following modification of the statement T^3:

(T^2) if $C(x)$ and $PC(x)$ and $E(x) - I(x) = 0$ and $m'(x) - M'(x) \neq 0$ and $ORG_A(x) - ORG_I(x) \neq 0$ and $mer(x) \neq 0$ and $ORG(x) - ORG_E(x) \neq 0$ and $D(x) - S(x) \neq 0$ and $Org(x) - ORG(x) \neq 0$,
then $P(x) = k(x) + k(x)p'(x) + mer(x) + R(x) + e(m'(x) - M'(x))$,

where the function e satisfies the following conditions (*ibid.*):

(1) $e(m'(x) - M'(x)) = 0$, if $(m'(x) = M'(x))$ or $(m'(x) > M'(x)$ and $ORG(x) = ORG_E(x))$ or $(m'(x) < M'(x)$ and $ORG(x) = ORG_E(x))$,

(2) $e(m'(x) - M'(x)) > 0$, if $(m'(x) > M'(x)$ and $ORG(x) < ORG_E(x))$ or $(m'(x) < M'(x)$ and $ORG(x) > ORG_E(x))$,

(3) $e(m'(x) - M'(x) < 0$, if $(m'(x) > M'(x)$ and $ORG(x) > ORG_E(x))$ or $(m'(x) < M'(x)$ and $ORG(x) < ORG_E(x))$,

and, moreover, in cases (2) and (3):

$$|\Delta e(m'(x) - M'(x))| < |\Delta pr(x)|,$$

where *pr* stands for profit (Marx (1), Vol. III, Chap. XI). The statement T^2 will be termed the law of prices of production under the consideration of merchant profit, rent and fluctuations of rate of suplus value.

Now, the statement T^2 contains two simplifying conditions p_1 and p_2. According to them, Marx in Vol. III of his *Capital* still considers ideal free-competition and closed economy. The presupposition p_1 assumes

> the removal of all monopolies (. . .). A great density of population is another requirement. The (. . .) condition implies the abolition of all laws preventing the labourers from transferring from one sphere of production to another (. . .); the greatest possible reduction of labour in all spheres of production to simple labour; the elimination of all vocational prejudices among labourers (Marx (1), Vol. III, pp. 192–193).

And, moreover, this ideal free-competition economy is, according to the assumption p_2, a closed one. But the removal of the conditions p_1 and p_2 would lead to further modifications of the law of value, if Marx was able to complete his work on *Capital*. As he wrote:

the law of value in its international application is yet more modified by this, that on the world-market the more productive national labour reckons also as the more intense, so long as the more productive nation is not compelled by competition to lower the selling price of its commodities to the level of their value. In proportion as capitalist production is developed in a country, in the same proportion do the national intensity and productivity of labour there rise above the international level. The different quantity of commodities of the same kind, produced in different countries in the same working-time, have, therefore, unequal international values, which are expressed in different prices, i.e., in sums of money varying according to international values (Marx (1), Vol. I, p. 560).

To sum up: the conditions p_1 and p_2 have not been removed because Marx was not able to complete his *Capital.* But, on the other hand, the statement T^2 has a quite large range of application – to all commodities produced in economies that are close to the free-competition ideal and to the closed economy. In other words, Marx most likely thought that his theoretical ideally free-competition and closed economy, for which T^2 holds, comes close enough to real economic conditions. For that law can be applied to real prices of a large amount of commodities with deviations that would not exceed "admissible" limits.

The statement T^2 applies, then, "approximately" to empirical commodities. But what does that really mean? Let us quote Engels' explanations in his *Supplement to Capital*:

the Marxian law of value holds generally, as far as economic laws are valid at all, for the whole period of simple commodity production, that is, up to the time when the latter suffers a modification through the appearance of the capitalist form of production. Up to that time prices gravitate towards the values fixed according to the Marxian law and oscillate around those values, so that *the more fully simple commodity production develops, the more the average prices over long periods uninterrupted by external violent disturbances coincide with values within a negligible margin* (Engels (1), p. 109). [Italics – L.N.]

Now, one may state that the statement T^2, i.e., the most developed form of the law of value, is "satisfied approximately" by empirical commodities, if they satisfy (in the strict sense) the following statement:

(AT^2) if $C(x)$ and $PC'(x)$ and $E(x) - I(x) \approx 0$ and $m'(x) - M'(x) \neq 0$ and $ORG_A(x) - ORG_I(x) \neq 0$ and $mer(x) \neq 0$ and $ORG(x) - ORG_E(x) \neq 0$ and $D(x) - S(x) \neq 0$ and $Org(x) - ORG(x) \neq 0$,
then $P(x) \approx k(x) + k(x)p'(x) + mer(x) + R(x) + e(m'(x) - M'(x))$,

where $PC'(x)$ means that a commodity x is produced in an economy that is an "almost free-competition" economy. The latter condition is a qualitative one, hence it is hard to express "approximation" to the condition in a more strict way as it is possible, for instance, in the case of the condition p_2.

The above analysis shows that in order to explain the formation of real prices of commodities, Marx resorts to a complicated procedure: he adopts many simplifying assumptions, formulates a simple hypothesis about the general conditions of the formation of prices under those assumptions and then he removes those assumptions correcting simultaneously the hypothesis in question and obtaining the sequence of its "transformed forms" each of them being based on a lesser amount of previously introduced simplifying conditions. The procedure ends with the "approximate application" of the last of those "transformed forms" of the initial hypothesis. This may be illustrated as in Figure 1.

$p_1, p_2, p_3, p_4, p_5, p_6, p_7, p_8$	T^8
$p_1, p_2, p_3, p_4, p_5, p_6, p_7, \bar{p}_8$	T^7
$p_1, p_2, p_3, p_4, p_5, p_6, \bar{p}_7, \bar{p}_8$	T^6
$p_1, p_2, p_3, p_4, p_5, \bar{p}_6, \bar{p}_7, \bar{p}_8$	T^5
$p_1, p_2. p_3, p_4, \bar{p}_5, \bar{p}_6, \bar{p}_7, \bar{p}_8$	T^4
$p_1, p_2, p_3, \bar{p}_4, \bar{p}_5, \bar{p}_6, \bar{p}_7, \bar{p}_8$	T^3
$p_1, p_2, \bar{p}_3, \bar{p}_4, \bar{p}_5, \bar{p}_6, \bar{p}_7, \bar{p}_8$	T^2
$\bar{p}_1, \bar{p}_2, \bar{p}_3, \bar{p}_4, \bar{p}_5, \bar{p}_6, \bar{p}_7, \bar{p}_8$	AT^2
(a)	(b)

Fig. 1. (a) the lists of simplifying assumptions adopted by Marx in subsequent stages of analysis; $\bar{p}_i$ means that p_i is removed, i.e., non-p_i is adopted. (b) the sequence of statements composing the theory of value: T^0 is the law of value, T^7, ..., T^2 are its 'transformed forms', AT^2 is the 'approximate form' of T^2.

Let us examine Marx's method of inquiry more generally:

(I) Marx introduces some assumptions of which he knows *a priori* to be false in empirical reality (e.g., he introduces the assumption that the demand and the supply of any commodity are balanced all the time, in spite of the contrary being the case with commodities – changes of demand and supply are an ordinary phenomenon in a free-competition economy).

(II) With the above assumptions in mind he proposes the formula revealing what the phenomena in question depend on. The formula is based, then, on assumptions that do not hold in empirical conditions.

(III) These counterfactual assumptions are then removed and the formula, i.e., the consequent of the law in question, is corrected correspondingly. The "transformed forms" of the initial law are thus obtained which deal with conditions that are less and less abstract (i.e., satisfy less and less counterfactual assumption) and, at the same time, those conditions come closer and closer to the empirical ones.

NOTES

[1]As Lange (1) has shown, formulations of this type are not correct from the point of view of the modern theory of decision.

[2]My italics will be marked with the initials 'L.N.' in this book.

[3]Translation is corrected here.

CHAPTER 2

THE METHOD OF IDEALIZATION AND CONCRETIZATION

1. INTRODUCTORY COMPARATIVE ANALYSIS

In the introduction to this book, I claimed that the reconstruction of the Marxian research practice will be a good starting point for the construction of the Marxist methodology, only in the case that Marx turns out to say something new from the standpoint of contemporary methodological schools. Now, it seems that it may be proved even on the basis of the indefinitely expressed characteristics of his method presented in the first chapter.

The common-sense concept of science cherishes an idea that whatever one may say about science, at least one thing is certain: science describes the reality we live in. But Marx's law of value T^8 does not describe the empirical phenomenon of the formation of prices; Marx, however, not only formulates this law, but even calls it the basic law for the capitalist economy.

Nevertheless, it is possible that the Marxian method is paradoxical from the viewpoint of common sense only, but is not in the least surprising from the point of view of modern philosophy of science.

Let us therefore examine the positivist definition of a law:

> for an unrestricted universal to be called a law a plausible requiremet is that the evidence for it is not known to coincide with its scope of predication. (Nagel (1), p. 63.)

The evidence for the law is made of *known* facts fulfilling the antecedent and consequent of the law and the scope of predication of the law is made of all the facts falling under its antecedent and consequent (*ibid.*). Applying, then, Nagel's requirement to the law of value T^8 we shall see that the set of facts which fulfill both the antecedent and the consequent of T^8 is empty, since there is no such commodity for which both

$$PC(x) \text{ and } E(x) - I(x) = 0 \text{ and } \ldots \text{ and } Org(x) - ORG(x) = 0,$$

and

$$P(x) = l(V(x)),$$

and the more so for the set of *known* facts which fall under this law. The evidence (in Nagel's sense) for the law of value not only fails to be the subset of its scope of predication but is identical to it, for both sets are empty. Therefore, the Marxian law of value is not a law in Nagel's sense at all or, more generally, in the positivist sense.

Let us now examine hypotheticism. The law of science is a statement which is, for Popper, physically necessary, i.e., holds true in all the worlds that, if they differ from our world at all, they do so only with respect to the initial conditions of that statement (Popper (1), p. 433). The initial conditions of the general statement

(1) for each x, if $A(x)$, then $B(x)$

are sentences

$$A(a),\ A(b),\ \ldots .$$

where a, b, ... are actual objects. The statement of the above form is a law if it holds true in every world which differs from ours at most in the initial conditions, but does not differ in fulfilling the consequent. The condition $B(x)$ is therefore valid in our world and in every other world which differs from ours only in the initial conditions. But the law of value T^8 does not fall even under this more subtle scheme. The dependence

(2) $P(x) = l(V(x))$

from the consequent of the law T^8 is not fulfilled in our world, for in the actual world prices do not depend on value alone; they depend, for instance, on the level of the organic composition of a given capital, on the changes of supply and demand, etc. (see Chapter 1, Section 2). Hence at most the following dependence could be fulfilled in "our world" (i.e., in the actual free-competition economy of Marx's time):

(3) $P(x) = k(x) + k(x)p'(x) + mer(x) + R(x) + e(m'(x) - M'(x))$;

the formula is the consequent of the statement T^2 (see Chapter 1, Section 2). Since the consequent of the law of value is not fulfilled in our world, the law is not a law in Popper's sense. The law of value, to use Popper's terminology, is fulfilled in the world which differs from ours in its nomological structure. That is why Marx never uses the law T^8 for a direct explanation of empirical phenomena, but "transforms" it establishing the statements T^7, T^6, . . . being fulfilled in the worlds which are nomologically closer and closer to our world.

To sum up. Marx conducts his inquiry in a way which is not only incompatible with the common sense view of science, but also in a way which is incompatible with the claims of modern philosophy of science: the law of value T^8 put forward as one of the main laws of *Capital* is neither a law in the positivist sense, nor a law in the hypotheticist sense. Buth of these influencial modern trends of philosophy of science fail to validate the scientific method of Marx.[1]

2. ANOTHER EXAMPLE: THE THEORY OF REPRODUCTION

Now the problem may be posed as to whether the theory of value is a typical piece of Marx's scientific conduct in *Capital.* It seems to me that the answer should be positive. I shall demonstrate it by the example of another fundamental theory of *Capital,* namely, the theory of reproduction.

The theory of reproduction is also based on numerous unrealistic assumptions of which only two will be mentioned:

q_1: the values of commodities are constant in the economy x

q_2: accumulation in the economy x equals zero.

Symbolically

$$q_1: \ Val(x) = \text{const}$$
$$q_2: \ M_a(x) = 0$$

The first of these assumptions Marx adopts stating that according to his presupposition no revolutionary changes in the values of the components of production capital occur, since these changes are generally interferences which can be understood only if they are considered as deviations from the unchanged value relations (Marx (1), Vol. II, Chap. 20, §I). Introducing the second assumption, Marx states that simple reproduction in an unchanged scale is an abstraction in so far that it cannot be assumed that no accumulation might occur under capitalist conditions, i.e., that there could be no enlarged reproduction (*ibid.*).

Therefore well known Marxian schemes of simple reproduction are:

$$P_1 = C_1 + V_1 + M_1$$
$$P_2 = C_2 + V_2 + M_2$$

where P denotes the global product of a given economy, C the constant capital, V the variable capital, M the surplus value, and where 1, 2, denote the indexes of the first and second sector of a given economy. The schemes are in force under assumptions q_1 and q_2. Therefore they may be presented in the following, more complete, form:

(R_1^2) if $E(x)$ and $Val(x) = \text{const}$ and $M_a(x) = 0$,
then $P_1(x) = C_1(x) + V_1(x) + M_1(x)$,

(R_2^2) if $E(x)$ and $Val(x) = \text{const}$ and $M_a(x) = 0$
then $P_2(x) = C_2(x) + V_2(x) + M_2(x)$.

From these two theses Marx draws his well-known condition of balance for simple reproduction. It also holds under the assumptions q_1 and q_2:

(c^2) if $E(x)$ and $Val(x) = \text{const}$ and $M_a(x) = 0$,
then $C_2(x) = V_1(x) + M_1(x)$

Statements (R_1^2), (R_2^2), (c^2) are in force under the assumptions:

q_0: $E(x)$ (x is a capitalist economy),
q_1: $Val(x) = \text{const}$,
q_2: $M_a(x) = 0$,

of which the first can, and the latter two cannot, be satisfied in the actual world. Condition q_0 corresponds, then, to condition p_0 from the law of value (see Chapter, Section. 1) and conditions q_1, q_2 are of the same nature as conditions p_1–p_8 from the law of value. The theses of the simple model of reproduction are therefore of the same character as the law of value.

The passage to the schemes of enlarged reproduction is performed in the same way as the passage from the law of value T^8 to its "transformed forms" T^7, T^6, etc. (see Chapter 1, Section 2). Namely, Marx removes the assumption q_2 and he takes realistically into account that capitalist economy x does accumulate, i.e., utilises a part of surplus value for an increase of the means of production (M^c) and for an employment of more labour force (M^v):

$$0 < M_a(x) = M^c(x) + M^v(x).$$

At the same time the consequents of the schemes (R_1^2), (R_2^2) are corrected with respect to the fund of accumulation M_a. In this way, the schemes of enlarged reproduction are obtained:

(R_1^1) if $E(x)$ and $Val(x) = \text{const}$ and $M_a(x) > 0$, then $P_1(x) = C_1(x) + V_1(x) + M_1(x) + M_1^c(x) + M_1^v(x)$;

(R_2^1) if $E(x)$ and $Val(x) = \text{const}$ and $M_a(x) > 0$,
then $P_2(x) = C_2(x) + V_2(x) + M_2(x) + M_2^c(x) + M_2^v(x)$.

From these theses the condition of balance for the enlarged reproduction follows:

(c^1) if $E(x)$ and $Val(x) = \text{const}$ and $M_a(x) > 0$,
then $C_2(x) + M_2^c(x) = V_1(x) + M_1^v(x) + M_1^c(x)$.

It is clear that the statement R_i^1 which accounts for accumulation is related to the appropriate statement R_i^2 abstracting from accumulation in the same way as the theorem T^7 is related to the law of value T^8, or the theorem T^6 is related to the theorem T^7, etc. (see Chapter 1, Section 2).

The structure of the theory of reproduction (in its part reconstructed here) may, then, be presented as shown in Figure 2:

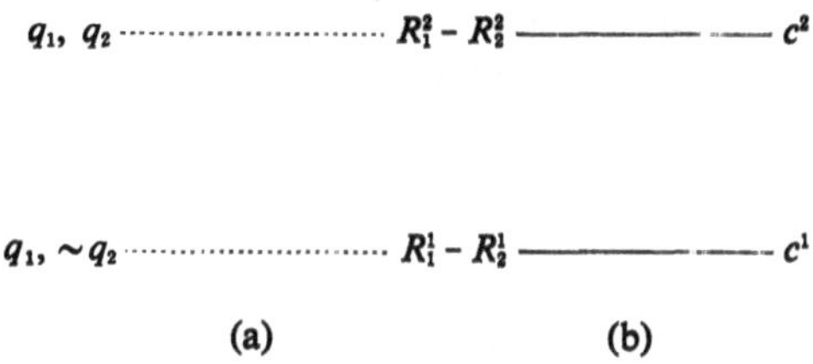

Fig. 2. (a) The lists of assumptions adopted by Marx in two stages of analysis (i.e., that of the model of simple reproduction and that of the model of enlarged reproduction). (b) The sequence of sets composing the theory of reproduction: each set contains two premises and their conclusion (the condition of balance for the simple or enlarged reproduction).

It can be seen that Marx, when constructing the theory of reproduction, followed the same method he employed in constructing the theory of value:

(I) he introduces unrealistic assumptions (e.g., conditions q_1 and q_2 postulating the economy with stable values and with zeroing accumulation);

(II) he establishes the basic laws with these assumptions in mind (so-called schemes of simple reproduction);

(III) he modifies these laws removing the assumptions in question and correcting the consequents of the laws (in this way, the schemes of enlarged reproduction are obtained).

This cannot be the work of accident: the two principal theories of *Capital* are built in the same way. Apparently, this is *the* Marxian method of building theories – in order to explain the investigated phenomena he introduces unrealistic assumptions, establishes the laws being in force under them and modifies the laws removing the assumptions and successively correcting the formulas (consequents) occurring in these laws.

3. SOME BASIC NOTIONS: THE METHOD OF IDEALIZATION AND CONCRETIZATION

I shall now try to conceptualize the Marxian method in some general notions. Let us assume that set U is the universe of discourse. A *realistic assumption* is the propositional function $G(x)$ which is fulfilled by any element of the universe U. For instance, 'x is a commodity' is a realistic assumption – the condition establishes the universe of discourse of the theory of value. Let us call the following propositional function an *idealizing assumption:*

(4) $p(x) = 0$

if and only if 0 symbolizes the minimum value of the magnitude p and for each actual object a: $p(a) \neq 0$. In this sense, idealizing conditions have an unrealistic nature – they are not fulfilled by any actual object. One may notice that conditions p_0 and q_0 are realistic assumptions provided by the theory of value and that of reproduction correspondingly. But conditions $p_1 - p_8$ and $q_1 - q_2$ are of a different nature – they are idealizing conditions. For instance,

$$q_2\colon\ M_a(x) = 0$$

falls under scheme (4) and satisfies required conditions. That is not so evident in all the cases in question. Condition q_1 has to be rewritten to fall under the schema (4) according to the equivalence: '$p(x)$ = const' meaning the same as '$\Delta p(x) = 0$' does. In other cases (e.g., p_1) the above definition of the idealizing condition can be applied only in an approximate way, i.e., the definition presents, so to speak, the pure form of an idealizing condition. Therefore it has to be enlarged to include some untypical cases. We shall discuss that more carefully in the third part of this book.

An *idealizational statement* will be a general statement of the following form:

(T^k) if $G(x)$ and $p_1(x) = 0$ and ... and $p_{k-1}(x) = 0$ and $p_k(x) = 0$, then $F(x) = f_k(H_1(x), \ldots, H_n(x))$,

where $G(x)$ is a realistic assumption while $p_1(x) = 0, \ldots, p_k(x) = 0$ are idealizing assumptions ($k > 0$). The idealizational statement determines in what way a given magnitude depends on another magnitude in idealized conditions, i.e., conditions fulfilling idealizing assumptions. Idealized conditions are defined with the aid of idealizing assumptions: a closed economy is defined as an economy where imports are equal to exports; a simply reproducing economy is defined as an economy where the fund of accumulation equals zero; etc. One may easily observe that the law of value T^8 is an idealizational statement with eight idealizing conditions. The same holds for the schemes of reproduction $R_1^2 - R_2^2$: they are idealizational statements equipped with two idealizing assumptions.

As has been said, the idealizational statement T^k is concerned with idealized conditions of the kth degree which differ from empirical conditions in kth respects shown in idealizing assumptions. In order to pass from the idealized conditions of that type to the conditions less idealized and hence to the conditions closer to empirical ones, one should remove the assumption

$$p_k(x) = 0,$$

i.e., replace it with the realistic condition

$$p_k(x) \neq 0,$$

and introduce the correction into the consequent of T^k expressing the influence of the magnitude p_k upon the magnitude in question F. Therefore the following statement will be called the first concretization of T^k with respect to the functions g, h:

(T^{k-1}) if $G(x)$ and $p_1(x) = 0$ and ... and $p_{k-1}(x) = 0$ and $p_k(x) \neq 0$, then $F(x) = f_{k-1}(H_1(x), \ldots, H_n(x), p_k(x)) = g[f_k(H_1(x), \ldots, H_n(x)), h(p_k(x))]$.

The function h expresses the impact of the factor p_k on the investigated magnitude F (*correctional function*), while g expresses the type of this impact, for example the fact that it constitutes a strengthening or weakening of the dependence f_k; that is why g is called the *directional function.* In the limiting case of *final concretization* all the idealizing assumptions are removed and all the appropriate corrections are introduced:

(T^0) if $G(x)$ and $p_1(x) \neq 0$, and ... and $p_k(x) \neq 0$,
then $F(x) = f_0(H_1(x), \ldots, H_n(x), p_k(x), \cdots, p_1(x))$
$= [nf_1(H_1(x), \cdots, H_n(x), p_k(x), \cdots, p_2(x)), m(p_1(x))]$.

In this extreme case, the influence of all the factors $p_k, \ldots, p_1$ upon the magnitude F is taken into account.

One may notice that the statement T^0 is not an idealizational statement at all: it contains only realistic conditions. A statement that is equipped only with realistic conditions will be called a *factual statement*. The degree of resemblance of the conditions considered in the successive stages of theory-construction increases, then, in the process of concretization from the minimum (in the case of initial idealizational statement) to the maximum, i.e., identity (in the case of the final concretization achieving the level of the factual statement).

The final concretization of a given idealizational statement is constructed, in fact, rather rarely, if at all. Scientists end their procedure of concretization at some point and instead assume that the influence of the remaining factors on the investigated magnitude is 'sufficiently small' hence the real values of that magnitude are 'sufficiently close' to its theoretical values, as established by the latest idealizational statement. Thus they establish a factual statement which says that if the given real conditions are 'sufficiently close' to those described by idealizing conditions, then the consequent of the statement is fulfilled with 'sufficient precision'. Thus, for instance, the following statement is an *approximation* of the idealizational statement T^{k-1}:

(AT^{k-1}) if $G(x)$ and $p_1(x) \leqslant \alpha_1$ and... and $p_{k-1}(x) \leqslant \alpha_{k-1}$ and $p_k(x) \neq 0$, then $F(x) \approx f_{k-1}(H_1(x), \ldots, H_n(x), p_k(x))$,

where α_i, ε are numbers (thresholds of approximation) established a priori by the researcher. Therefore, the usual way of testing idealizational statements is the following: the initial idealizational statement T^k is approximated, i.e., its approximation AT^k which is a factual statement is established; if the results of observations concerning the magnitude F agree with the theoretical value of F (within the interval defined by ε), the statement AT^k, and hence T^k, are accepted; if not, the concretization of T^k follows and the statement T^{k-1} is established which is in turn approximated and the factual statement AT^{k-1} is considered; if the results of observations concerning F agree with the theoretical values of F (within the interval defined by ε), the statement AT^{k-1}, and hence T^{k-1} and T^k are accepted; if not, the concretization of T^{k-1} follows resulting in the statement T^{k-2} which is in turn approximated, etc. If the procedure does not result in achieving an agreement with observed facts, the statement T^k is falsified. These matters will be considered more carefully in the third part of the book.

What is important now is that these general schemes include (within the usual 'interval of tolerance' accepted in methodological researches) reconstructed above (see Chapter 1 and Chapter 2, Section 2) pieces of Marx's scientific practice in *Capital*. The relation which holds between T^7 and T^8 can be interpreted as the relation of concretization: T^7 is a less abstract idealizational theorem than T^8 and takes into account the influence of some factor (uneveness of the organic composition of capital within a given branch of production) which is neglected in T^8 due to the assumption p_8. Similarly, in other cases: T^6 is a concretization of T^7 etc, and also the theses of enlarged reproduction are concretizations of the schemes of simple reproduction (see Section 2). In the case of the theory of value the relation of approximation is also applied. The statement T^2 which is an idealizational one is approximated – due to that the statement AT^2 of the factual nature is accepted within the theory in question.

To sum it up: one may state that the Marxian procedure of theory-construction in *Capital* falls, at least approximately, under the general schema of the method of idealization and concretization. With the aid of notions constructed so far Marx's method may be described in the following way:

(I) Marx introduces idealizing assumptions,
(II) next he establishes the idealizational law,
(III) and he concretizes it gradually or approximates it.

4. CURRENT INTERPRETATION OF THE MARXIAN METHOD

Interpretation of the Marxian method in terms of idealization is, undoubtedly, simply a proposal which, as with every interpretation, can be either corrected or rejected. At the very least the interpretation has the virtue of being incompatible with the current interpretation of the

Marxist and, in particular, of the Marxian method. According to the latter interpretation

abstraction consists in induction. The starting point is (...) a comparative observation of an actual process in different historical conditions. (...) [Finally] that results in (...) formulation of the abstract economic categories, theoretical models and their corresponding laws and theories. All this happens by means of inductive reasoning. (Lange (1), pp. 176–177.)

Similarly the Russian philosopher M. Rozental states:

Induction plays the basic role in motion of thought from concrete to abstract (...), due to it one can reach the simplest abstractions being a starting point for gradual approximation to concretes. (Rozental (1), p. 495.)

According to this view he treats Marx's law of value in *Capital* as an inductive generalization of numerous economic facts (*ibid.*, p. 495). And concretization is conceived within the current interpretation of the Marxist method as deduction of more specific statements from previously established inductive generalizations. Hence concretization leads to less general statements describing correlations and connections between less general aspects of phenomena:

gradual concretization consists in deduction but induction plays, however, an important additional role. The theoretical model is gradually complemented with additional more specific assumptions and after that, by means of deductive reasoning, the concretization of categories laws and (...) theories follows. (Lange (1), p. 177.)

The main failure of the current interpretation of the Marxist method is the following: there are no grounds, accepting this view, to differentiate between the Marxist and positivist methodology (in the extreme version of the latter which has been rather rejected in modern positivism). In fact, it was J. S. Mill who reduced the concept of scientific law to that of strictly general statement and who saw the predominant role of induction in empirical sciences. In this connection, it is not surprising that followers of the discussed interpretation of the Marxist methodology do not say about the scientific method any more than what J.S. Mill did. Their image of the method is exactly the same: observation, inductive generalization, deduction. The only difference is that Mill failed to use the Marxist phraseology: induction was called by him – induction, and deduction – deduction, while modern Marxists use Marx's terms speaking of abstraction and concretization...

One may easily notice that the current interpretation of the Marxian method is entirely inadequate. The law of value T^8:

(T^8) if $C(x)$ and $PC(x)$ and $E(x) - I(x) = 0$ and $m'(x) - M'(x) = 0$ and $ORG_A(x) - ORG_I(x) = 0$ and $mer(x) = 0$ and $ORG(x) - ORG_E(x) = 0$ and $D(x) - S(x) = 0$ and $Org(x) - ORG(x) = 0$,
then $P(x) = l(V(x))$

cannot be justified inductively because of the simple reason that there is no such commodity which fulfills assumptions occurring in the antecedent of the law. In other words, for each actual commodity a, the conjunction

$$PC(a) \text{ and } E(a) - I(a) = 0 \text{ and} \ldots \text{and } Org(a) - ORG(a) = 0$$

is false, and therefore the law of value T^8 cannot be introduced by means of the inductive reasoning:

$$\frac{\begin{array}{l} P(a) \text{ and } Q(a) \\ P(b) \text{ and } Q(b) \\ \ldots\ldots\ldots\ldots\ldots \\ P(n) \text{ and } Q(n) \end{array}}{\text{for each } x, \text{ if } P(x), \text{ then } Q(x)}$$

This pattern fails in the case of all idealizational statements (where the condition $P(x)$ contains idealizing assumptions), since the first element of each premise is false. Thus the interpretation of Marx's abstraction as induction seems to be quite misleading. It is rather interpretation of Marx's abstraction in terms of idealization which seems to agree with the following characteristic of the method of *Capital*:

In Book I we analysed the phenomena which constitute the *process of capitalist production* as such, as the immediate productive process, with no regard to any of the secondary effects of outside influences. But this immediate process of production does not exhaust the life span of capital. It is supplemented in the actual world by the process of circulation, which was the object of study in Book II (...). Considering what this third book treats, (...) it must locate and and describe the concrete forms which grow out of the *movement of capital* as a whole. (...) The various forms of capital, as evolved in this book, thus approach step by step the form which they assume on the surface of society, in the action of different capitals upon one another, in competition, and in ordinary consciousness of the agents of production themselves. (Marx (1), Vol. III, p. 25).

One may even say more: Marx criticizes other economists (e.g., Ricardo) precisely for his acceptance in his research practice of the notion of abstraction which is identical to the notion of abstraction as inducticn, i.e., to that notion of abstraction that is nowadays ascribed to Marx by many Marxists. Marx states:

One can see that in this first chapter [of Ricardo's *Principles*–L.N.] not only are commodities assumed to exist – and when considering values as such, nothing further is required – but also wages, capital, profit, the general rate of profit and even, as we shall see, the various forms of capital as they arise from the process of circulation. (Marx (5), Part II, p. 168).

Seeing Ricardo's inconsistencies

The vulgar mob has therefore concluded that theoretical truths are abstractions which are at variance with reality, instead of seeing, on the contrary, that Ricardo does not carry true abstract thinking far enough and is therefore driven into false abstractions. (*Ibid.*, p. 437).

According to Marx, then, Ricardo is wrong because of his inability to idealize investigated phenomena, i.e., to abstract from secondary influences: Ricardo applies the method of induction but he does not apply the method of abstraction in the Marxian sense, i.e., the method of idealization.

5. GALILEO OF THE SOCIAL SCIENCES

The method of idealization is not the only universal method applied in empirical sciences from its origin. In every empirical science one may distinguish two basic periods: one may be called the period of premature science and the other merits the name of the period of theoretical science. In the first period a science applies, in principal, inductive methods. After collecting an empirical knowledge about a given domain there occurs the possibility at building a theory explaining those empirical generalizations. The first theory is built by means of the method of idealization. In each empirical science that moment in its history is marked as the methodological breakthrough. My thesis is that the breakthrough in political economy was accomplished by Karl Marx. To demonstrate this I shall compare it with Galileo's revolution in physics.

The revolution consisted in making evident the false nature of the world image which senses reveal. We only see those phenomena which are the results of what is essential and also of what is less important in reality. The aim of science consists in revealing how these phenomena depend on principal factors. The discrepancy between scientific laws and observation is unavoidable, but one can minimize it by taking secondary factors into account. When doing the latter science opposes common sense which thinks evident only what is observed. Galileo says:

experiences which clearly state against the annual movement are seemingly so contrary to the theory that (...) I cannot find the words to express my admiration for Aristarches and Copernicus who managed to put reason into a frame which forced the senses to withdraw their trust in the apparent meaning of sensory data (Galileo (1), pp. 353–354).

This proves how great is

the elevation of these minds which accepted these views and took them as true ones overcoming the testimony of their own senses with the quickness of minds and prefering that which reason dictated to what senses and experiments seemed to offer (*ibid.*, p. 354).

In order to establish the true image of phenomena independent of the appearances resulting from senses one has to apply the method of idealization: try to imagine what the investigated phenomenon would look like if the disturbing factors did not occur in reality. As Galileo writes

The philosopher-geometrician who wants to investigate in reality what has abstractly been proved should exclude the interfering influences of the matter from the calculations. (*Ibid.*, p. 225.)

In order to stress that point I shall compare the frame within which the same phenomenon described in their laws of motion is considered in the physics of Aristotle and that of Galileo.

According to Aristotle, the observation of moving bodies shows us that two factors influence their movement: the external force and the resistance of the environment within which the bodies occur. In order for motion to occur the external force has to prevail over the resistance. When the force is equal to the resistance, movement does not occur. The question as to how would the body behave if no resistance occurred (e.g., in a vacuum) was answered by Aristotle in a common sense way. No one need worry about it, since "one should first explain what a vacuum among the moving bodies is. For no one has noticed it within the world so far" (Aristotle (1), p. 125). Galileo knew well enough that a vacuum could not be perceived. But this did not prevent him from asking the question that did not make sense in Aristotle's physics: how would the body move if the environment did not interfere? For instance, how would a sphere move on a plane, if "we did not account for the resistance of air interfering with the sphere piercing its way through it?" (Galileo (1), p. 155). The sphere in question is a "perfectly round one, so that all external and chance obstacles are removed" (*ibid.*, p. 155). Particularly, the resistance of air is a minor element since "it disturbs all the movements in an infinitely various set of ways, according to the changes in shape, weight and the velocity of the bodies (...). One cannot give the strict theory for all those infinitely diverse cases of weight, velocity and shape. One has to abstract from them in order to treat the subject in a scientific way, and the conclusions found and proved without the interferences may be used within the framework provided by the experiments" (Galileo (2), pp. 181–182).

Galileo's answer to his new problem was the principle of inertia in its first and – as it was later evident – imprecise form:

(5) if the moving body is an ideal sphere rolling on an ideally smooth plane, then, when no environmental resistance is found, its movement is constant and steady movement on the horizontal plane (*ibid.*, p. 176).

The principle of inertia (5) refers directly to the ideal case of movement which does not occur in reality. One may see that this principle is an idealizational statement. Its approximation is the following factual theorem:

(6) if the moving body is sufficiently close to the ideal sphere and rolls along the short, sufficiently smooth surface, and the resistance of the environment is small, then its movement is an uniform movement along the straight line (i.e., it coincides with going in circles for small distances).

The Theorem (6) allows prediction of actual movements of real bodies and can be experimentally verified. And, in fact, Galileo had tested it in many ways. The methodological difference between Aristotelian and Galilean physics is now apparent as is the nature of the methodological breakthrough in physics which is commonly associated with Galileo's name. Aristotle's theory of movement was a common sense one – it tried to describe the phenomena as they appear to us in everyday life. The method was to observe and to systematize observations in order to generalize them. The inductive generalizations of the Aristotelian physics were surprisingly well confirmed by everyday experience: everybody could confirm them using his own observations. The breakthrough of Galileo consisted in the discovery that observation alone cannot suggest the contents of theories. For although we observe the results of what is essential and what is less significant, it is necessary to introduce the simplifying assumptions which allow us to eliminate factors of minor importance. That is why science establishes idealizational laws and modifies them, taking into account the influence of previously disregarded factors, i.e., concretizes or approximates idealizational laws. In brief, the Galilean breakthrough consisted in the introduction of the method of idealization in physics.

Historically speaking, Marx was in a situation resembling that of Galileo. The political economy of the period he lived in was predominantly the one he had himself termed "vulgar economy". Here is the flat description of the orientation provided by Marx:

All that vulgar economy wants to do and, in fact, does is that it describes, explains and apologetically synthetizes the images of the agents of bourgeois production, who remain within the power of the relations of this production. No wonder it feels at home in the external form of appearance of these relations, which are there senseless and prima facie full of contradictions – which is natural since *science would be superfluous if the form of appearance of things and their essence were simply identical.* The more these relations seem to be self-explanatory the more their internal connections are hidden from vulgar economy, although they appear as simple in current imaginary. (Marx (1), Vol. III, p. 825) [italics – L.N.]

Evident here is the Marxian argument against vulgar economy – it provides the common-sense knowledge based on generalization of observations; it systematizes only what is available for the commonsense consciousness. This is not a proper method, if we do not want a simple description of phenomena, but their explanation. What we are able to observe is the joint result of principal and secondary factors. Observation itself does not enable us to distinguish between principal and secondary factors influencing a given phenomenon, hence

the tracing of an apparent movement which occurs on the surface of the phenomena to the internal, real one must be the work of science. (Marx (1), Vol. III, p. 325)

According to the proper scientific method, an investigator should separate the principal and secondary factors for a given phenomenon and abstract from them establishing the law connecting the phenomenon with principal determinants of it. Thus idealizational laws express scientists' convictions concerning the essential stratification of the world. It is only afterwards that the modifications caused by the minor factors should be considered. The second phase being concretization of idealizational laws which express scientists convictions concerning the influences of secondary elements of the world upon its real nature revealed in initial idealizational laws.

This connection between the method of idealization and essentialist presuppositions of science will be discussed later (see Part III of this book). Let us notice, however, that from Marx's point of view the real justification for applying the method of idealization is the objective essential differentiation of the world. That is why Marx strongly distinguishes between "two methods of investigation":

One of these conceptions fathoms the inner connection, the physiology, so to speak, of the bourgeois system, whereas the other takes the external phenomena of life, as they seem and appear and merely describes, catalogues, recounts and arranges them under formal definitions (Marx (5), Part II, p. 165).

The first method is his own – it is the method of abstraction (idealization – as I interpret it) which is able to trace "the intrinsic connection existing between economic categories or the obscure structure of the bourgeois economic system" (*ibid.*, p. 165). The second one was justified at the beginning of the development of political economy, e.g., in Smith's conception, but it is backward nowadays (i.e., in Marx's time) when what is needed is an explanatory theory revealing the true nature of economic phenomena. A theory of that type may be built only idealizationally. This is the main task of *Capital*: to build an idealizational theory of the bourgeois economy. And, at the same time, to introduce the method of idealization into the social sciences. To be the Galileo of the social sciences.

NOTE

[1]It does not imply that the method of idealization has not been noticed in methodology. On the contrary, many valuable observations and suggestions have been proposed (see, e.g., Hempel (1), Watkins (1), Bunge (1), Rudner (1)). What I mean is that this method has not been treated as the basic method of science. On the contrary – it has been ignored or interpreted in terms of other methods.

PART II

MARX'S IDEAS AND MODERN METHODOLOGICAL TRENDS

The goal of this part of the book is to compare Marx's method of idealization (to be strict: Marx's method of abstraction interpreted in terms of idealization) with well known and influential modern methodological trends or schools. It seems to me that they are of two kinds: those that belong to the tradition of the German philosophy of humanities and those that belong to the empiricist tradition. We shall distinguish within the first group the Weberian methodological doctrine and within the second group two trends will be discussed: positivism and hypotheticism. At the end some of the newest 'anti-empiricist' trends in the philosophy of science will be compared with what we reconstructed from the Marxian works and what is presented here as an admissible interpretation of the methodology of the author of *Capital.*

CHAPTER 3

IDEALIZATION AND IDEAL-TYPICAL METHOD: MARX AND WEBER

1. IDEAL-TYPICAL METHOD

According to Max Weber

> . . . all specifically Marxian 'laws' and developmental constructs – insofar as they are theoretically sound – are ideal types (Weber (1), p. 103).

My aim here is to prove that this opinion – which ascribes to Marx the role of a scientist who unconsciously accepted the Weberian conception of ideal types – is false, since there is an important difference between Marx's abstraction and Weber's ideal-typical procedure.

Here is Weber's characteristics of ideal types:

> An ideal type is formed by the one-sided *accentuation* of one or more points of view and by synthesis of a great many diffuse, discrete, more or less present and occasionally absent *concrete individual* phenomena, which are arranged according to those one-sidedly emphasised viewpoints into a unified *analytical* construct. In its conceptual purity, this mental construct cannot be found empirically anywhere in reality. It is a *Utopia*. (*Ibid.*, p. 90.)

Thus, for instance, economists use the ideal type of rational agent, i.e., someone whose actions are performed

> . . . in the absence of physical and psychic 'coercion', emotional 'affects' and accidental disturbances of the clarity of judgement, in which we pursue a clearly perceived end by 'means' which are the most adequate in accordance with the extent of our knowledge. (*Ibid.*, pp. 124–125.)

The concepts denoting ideal types conceived of in this way are ideal-typical concepts. They are peculiar for the humanities. In natural sciences the only type of concepts are generic ones which

> . . . merely summarize the common features of certain empirical phenomena (*ibid.*, p. 100).

thus denoting simply classes of these phenomena. As opposed to the generic notions the ideal-typical concept is a

. . . mental construct for the scrutiny and systematic characterization of individual concrete patterns which are significant in their uniqueness, such as Christianity, capitalism, etc. (*Ibid.*, p. 100.)

Let us discuss the notion of Christianity:

. . . empirically it exists in the minds of an indefinite and constantly changing mass of individuals and assumes in their minds the most multifarious nuances of form and content, clarity and meaning. Those elements of the spiritual life of the individuals living in a certain epoch of the Middle Ages, for example, which we may designate as the 'Christianity' of those individuals, would, if they could be completely portrayed, naturally constitute a chaos of infinitely differentiated and highly contradictory complexes of ideas and feelings. (...) If we raise the question as to what in this chaos was the 'Christianity' of the Middle Ages (which we must nonetheless use as a stable concept) (...) we see that here too in every individual case, we are applying a purely analytical construct created by ourselves. It is a combination of articles of faith, norms from church law and custom, maxims of conduct, and countless concrete interrelationships which we have fused into an 'idea'. It is a synthesis which we could succeed in attaining with consistency without the application of ideal-type concepts. (*Ibid.*, p. 96.)

The above-mentioned difference between the humanities and natural sciences is not, in Weber's views, conditioned by the fact that the humanities investigate some other "kind of being" than the natural sciences. Since there is nothing in the objective nature of phenomena which would presuppose the employment of generic concepts to the analysis of the natural phenomena and the employment of ideal-typical concept to the analysis of the cultural phenomena. It is particularly not that

. . . as is often maintained, . . . cultural or psychic events for instance are 'objectively' less governed by laws (*ibid.*, p. 80).

The difference between these two groups of sciences stems from their different role in the human culture. The subject-matter of all empirical sciences is a chaotic multiplicity of phenomena. The task of science consists in ordering the chaos. The basic principle of ordering phenomena in the natural sciences is their classification into classes containing elements possessing some common features. Such a classification enables natural sciences to discover laws claiming that phenomena of a definite class cause in some circumstances phenomena of a different kind. In cultural sciences another manner of ordering phenomena is employed:

The transcendental presupposition of every *cultural science* lies not in our finding a certain culture or any 'culture' in general to be *valuable* but rather in the fact that we are *cultural beings*, endowed with the capacity and the will to take a deliberate attitude towards the world and to lend it *significance*. Whatever this significance may be, it will lead us to judge certain phenomena of human existence in its light and to respond to them as being (positively or negatively) meaningful. Whatever may be the content of this attitude – these phenomena have cultural significance for us and on this significance alone rests its scientific interest. (*Ibid.*, p. 81.)

Thus, although the whole science is a cultural product, only the cultural sciences are in such a situation in which the culture imposes upon them their cognitive problems: a humanist is interested in those phenomena only which have meaning assigned to them by the culture which he himself is acting. In order to investigate a fact of a cultural nature, i.e., facts being endowed with cultural meaning, one cannot use ordinary, i.e., generic notions. In order to accomplish that one has to resort to ideal-typical notions:

Social science in our sense is concerned with practical *significance*. This significance however can very often be brought unambigously to mind only by relating the empirical data to an ideal limiting case. (*Ibid.*, p. 94.)

2. IDEALIZATION AND THE IDEAL-TYPICAL METHOD: SCOPE

According to Weber's views the ideal-typical method is then used only within the cultural sciences. It is only in the cultural sciences that the method is an indispensable means of revealing the sense of social phenomena, i.e., their significance. As opposed to this in natural sciences,

Only those aspects of phenomena which were involved in the 'laws' could be essential from the scientific point of view, and concrete 'individual' events could be considered only as 'types,' i.e., as representative illustrations of laws. (*Ibid.*, p. 86.)

It is clear that Weber's view on the natural sciences is strongly influenced by the positivist philosophy of science. Weber, as was the case with all the anti-naturalists, was an antipositivist only in so far as methodological nature of the humanities was concerned; as for the natural sciences he admitted in advance that the positivist methodology was correct in characterizing them.

The makers of Marxism were assuming a quite different attitude, for they considered the method of idealization to be the principal method both in the natural sciences and in the humanities. For instance, Engels on the one hand stressed the idealizational nature of the Marxian economy, and on the other hand showed on the application of the method of idealization in the natural sciences. He wrote about the law of value:

According to the economists' assumption, only equal values are exchanged, and *in the sphere of abstract theory this is correct.* (Engels (1), p. 14) [italics – L.N.].

Writing about "the law of value and the distribution of the surplus value according to the rate of profit" he stressed that

Both attain their most complete approximate realization only on the presupposition that capitalist production has been everywhere completely established, i.e., that society has been reduced to the modern classes of landowners, capitalists (industrialists and merchants) and workers – all intermediate stages having been eliminated. This condition does not yet exist even in England. (Engels (2), p. 458.)

The same is concerned with basic notions and theorems of historical materialism:

Did feudalism ever correspond to its concept? Founded in the kingdom of the West Franks, further developed in Normandy by the Norwegian conquerors, its formation continued by the French Norsemen in England and Southern Italy, it came nearest to its concept – in the ephemeral kingdom of Jérusalem, which in the Assines de Jérusalem left behind it the most classic expression of the feudal order. Was this order a fiction because in a really classical form it achieved only in Palestine a shortlived existence, and even that, for the most part, on paper only? (*Ibid.*, pp. 458–459.)

One may see, then, that according to the founders of Marxism, idealization is a general procedure common to the social and natural sciences as well. It is a quite general thesis that Engels formulated in his discussion with C. Schmidt:

The objections you raise to the law of value apply to all concepts, regarded from the standpoint of reality. The identity of thinking and being, to use Hegelian language, everywhere coincides with your example of the circle and the polygon. Or the two of them, the concept of a thing and its reality, run side by side like two asymptotes, always approaching each other yet never meeting. . . . Because a concept has the essential nature of the concept and does not therefore *prima facie* directly coincide with reality, from which it had to be abstracted in the first place, it is neverthelses more than a fiction, unless you declare that all the results of thought are fictions because reality corresponds to them only very curcuitously, and even then approaching it only asymptotically. (Engels (2), p. 457.)

3. MARX AND ENGELS AS METHODOLOGICAL NATURALISTS

Now is an appropriate place to discuss the views of the founders of Marxism on the problem of the peculiar characteristics of the humanities.

If we agree that the principal characteristics of science is, according to Marx and Engels, the application of the method of idealization, then our previous considerations lead to the following *thesis of methodological naturalism:*

(1) The principal methodological properties of science concern both the natural sciences and the humanities.

This thesis can be ascribed, in my opinion, to the founders of Marxism. Indubitably: this is only some interpretation of their views, but it seems to me that there are good reasons (see Chapter 1 and Section 2 of this chapter) to ascribe it to them. The methodological doctrine of Marxism is often interpreted as anti-naturalist (e.g., Lukács (1), Korsch (1)), but revealing that the basic method of science according to Marx and Engels is that of idealization allows their methodological attitude to be interpreted in the naturalist manner.

To avoid misunderstandings one has to distinguish three other theses from Thesis (1):

(2) All presently (here: in Marx's time) formulated laws of the humanities may be drawn from the presently (here: in Marx's time) formulated laws of the natural sciences.

(3) All the presently (here: in Marx's time) formulated laws of the humanities may be drawn from the possible laws of the natural sciences.

(4) The conceptual apparatus of the humanities may be defined in terms of the (present) concepts of the natural sciences.

Accepting Thesis (1), the creators of Marxism rejected Theses (2) and (4) and did not decide about Thesis (3).

A number of arguments may be found that speak in favour of the supposition that Marx and Engels rejected the *weak thesis of ontological naturalism*. i.e., Thesis (2). Engels, for instance, says:

The essential difference between human and animal society consists in the fact that animals at most *collect* while men *produce*. This sole but cardinal difference alone precludes the simple transfer of laws of animal societies to human societies (Engels (4), p. 283.)

Engels had *a fortiori* the same views as regard the laws of chemistry, physics, etc. especially as he criticized the mechanistic materialists, among other things, for

die Anmaßung, die Naturtheorien auf die Gesellschaft anzuwenden (Engels (3), p. 215).

In other words, laws of the social sciences cannot be explained in terms of laws of the natural sciences as known to Engels. This does not, however, determine the negative answer to the question whether, in future, the explanation of laws concerning society will be possible in terms of the laws of the natural sciences. It is also difficult to find a positive answer to this question in the works of the creators of Marxism – that mostly depends on the interpretation of Engels' notion of the form of motion of matter and the relationship between them. That is why we leave open the question whether they accepted the *strong thesis of ontological naturalism* (3). As far as the *thesis of conceptual naturalism* is concerned, one may state that the creators of Marxism rejected it (at least in the weak form). Marx denied the possibility of reducing the basic concepts of political economy to that of the natural sciences. For instance, commodity is not the same as an object possessing definite physical properties:

the bodies of commodities, are combinations of two elements – matter and labour. If we take away the useful labour expended upon them, a material substratum is always left, which is furnished by Nature without the help of man (Marx (1), Vol. I, p. 43.)

but

As values, commodities are social magnitudes, i.e., something absolutely different from their properties as things. As values they reflect not only the relations between people in their productive activity. Value really implies exchanges, but exchanges are exchanges of things between men, exchanges which do not concern the things as such. (Marx (5), Part III, p. 149.)

Therefore, one must not say that

. . . exchange value of things is defined by their properties as things, that it is in general a natural property of these things, [for] No student of nature has discovered so far thanks to what natural properties tobacco and paintings are equivalent of each other in some defined proportion (*ibid.*, p. 150.)

This view is based on the assumption that

Everything that is not the result of human work is nature and as such is not a social value (*ibid.*, p. 499).

In order to understand properly the sense of this assumption one has to realize that for Marx work is a purposeful activity:

We pre-suppose labour in a form that stamps it as exclusively human. A spider conducts operations that resemble those of a weaver, and a bee puts to shame many an architect in the construction of her cells. But what distinguishes the worst architect from the best of bee is this, that the architect raises his structure in imagination before he erects it in reality. At the end of every labour-process, we get a result that already existed in the imagination of the labourer as its commencement. He not only effects a change of form in the material on which he works, but he also realizes a purpose of his own that gives the law to his *modus operandi*, and to which he must subordinate his will. (Marx (1), Vol. I, p. 178.)

The conclusion one might draw from the above is that the reason for which the economic concepts cannot be reduced to the concepts of natural sciences is that they presuppose the notion of purposeful activity which should be considered as one of the initial concepts of the social sciences (Kmita, Nowak (1) and (2), Kmita (1)).

It is, I think, clear from the above considerations that one difference between the method of idealization, that I ascribe to the founders of Marxism interpreting in this way their idea of abstraction, and the Weberian ideal-typical method consists in the different range of applicability of the methods in question. The former is to be a universal method applied in the social and natural sciences as well, while the latter is supposed to justify the exceptional position of the social ("cultural") sciences within science. In other words, Marx's and Engels' ideas may be interpreted in terms of methodological naturalism (see Thesis (1)), whereas Weber's concepts seem to be anti-naturalist (in the sense of (1)) ones.

4. IDEALIZATION AND THE IDEAL-TYPICAL METHOD: METHODOLOGICAL NATURE

Let us presently examine the problem whether idealization in Marx's sense and the method of idea types in the sense of Weber mean the same from the point of view of their methodological nature. The problems of methodological characteristics of ideal-typical statements were not, however, analysed by Weber, hence we have to reconstruct his views on the subject in the same manner we had previously reconstructed Marx's idea concerning similar matters.

Let us start from referring to Weber's instances of the application of the method of ideal types.

One can, for example, arrive at the theoretical conclusion that in a society which is organized on *strict* 'handicraft' principles, the only source of capital accumulation can be ground rent. From this perhaps, one can – for the correctness of the construct is not in question here – construct a pure ideal picture of the shift, conditioned by certain specific factors – e.g., limited land, increasing population, influx of precious metals, rationalisation of the conduct of life – from a handicraft to a capitalist economic organization. Whether the empirical-historical course of development was actually identical with the constructed one, can be investigated only by using this construct as a heuristic device for the comparison of the ideal type and the 'facts'. If the ideal type were 'correctly' constructed and the actual course of events *did not* correspond to that predicted by the ideal type, the hypothesis that medieval society was *not* in certain respects a *strictly* 'handicraft' type of society would be proved. And if the ideal type were constructed in a heuristically 'ideal' way – whether and in what way this could occur in our example will be entirely disregarded here – it will guide the investigation into a path leading to a more precise understanding of the non-handicraft components of medieval society in their peculiar characteristics and their historical significance. *If* it leads to this result, it fulfils its logical purpose, even though, in doing so, it demonstrates its divergence from reality. It was – in this case – the test of an hypothesis. (Weber (1), pp. 101–102.)

From the examples of the kind the following conclusion can be drawn with respect to the methodological nature of the Weberian method of ideal types:

(1) The statements which refer to ideal types (ideal-typical statements) are not synthetical but analytical statements, since they cannot be tested at all.

(2) The ideal-typical statement can deviate from reality but does not have to; in other words, it may happen that the ideal type agrees with the investigated phenomenon.

(3) The question whether the ideal-typical statement deviates from reality may be decided in the following way: one has to formulate the hypothesis that a definite ideal type coincides with the actual phenomenon, and from this and the ideal-typical statement as well empirical predictions may be deduced. The hypothesis in question is a synthetical statement, hence truthfulness of those predictions confirms it and their falsity falsifies it.

(4) If the ideal-typical statement agrees with the actual phenomenon, it can be employed in explanation and prediction in a regular manner – as generic statements (i.e., statements containing only generic notions).

(5) If the ideal-typical statement does not correspond to the actual phenomenon (which is usually the case), then it is helpful in establishing the degree of deviation between the appropriate ideal type defined by it and the actual phenomenon under investigation. Therefore, the statement stimulates research aiming at the explanation of the discrepancy. However, what is explained (a sentence about those features of the actual phenomenon that make it different from the ideal type in question) and what is used in explanation (sentences about causes of the deviation) are synthetical sentences.

As conclusion (1) states, the Weberian ideal-typical statements and the Marxian "abstractions" (interpreted as the idealizational statements) are two different types of sentences. The latter are synthetical statements, since they are able to be controlled on the basis of experience (with the aid of concretization and approximation). Whereas the former are analytical sentences – they cannot be subjected to empirical control. What is empirically testified are hypotheses about applicability of ideal-typical statements. Therefore, assuming "analytical-synthetical dualism" at least in the typological sense, none of the ideal-typical statements can be identified with an idealizational statement. The explication of those intuitions would be the following: an ideal-typical statement is the statement of the form:

(5) if $A_1(x)$ and ... and $A_m(x)$, then $B(x)$

such that it is not excluded that $A_i(x)$ is fulfilled by actual objects and the feature B is ascribed *ex definitione* to every object possessing the features $A_1, \ldots, A_m$. According to such an explication (5) is an analytical statement that may be non-vacuously satisfied. That the ideal-typical statement deviates from reality means that investigated phenomena do not satisfy its antecedent; if this is not the case, the statement agrees with reality. In the former case there is no guarantee that respective actual objects have the property B, while in the latter case there is such a guarantee thanks to the definition of the concepts $A_1, \ldots, A_m$.

Thus the Marxian "abstractions" (interpreted in terms of idealization) and the Weberian ideal-typical statements differ with respect to their logical status. This fact has some further important consequences.

Let us compare Weber's and Marx's ideas concerning explanation. Weber's model of explanation covers two subcases:

(1) When assumptions of the ideal-typical statement are satisfied (i.e., the hypothesis about applicability of the statement in question is true).

(2) When these assumptions are not fulfilled (i.e., the hypothesis in question is false).

In the first case the explained statement follows from the ideal-typical statement and the hypothesis about the applicability of the latter in a given case. That is:

(6)
$$\begin{array}{ll} \text{(1) if } A_1(x) \text{ and } \ldots \text{ and } A_m(x), \text{ then } B(x) & (t) \\ \text{(2) } A_1(a) \text{ and } \ldots \text{ and } A_m(a) & (h) \\ \hline \text{(3) } B(a). & \end{array}$$

In this schema, statement (1) is an ideal-typical statement, statement (2) is a hypothesis about applicability of (1) to a given object a and statement (3) is an explained statement.

In case (2) the model of explanation is of a different form. Let us assume that the hypothesis (2) turned out to be false, for it has been found that the object a has a property which is incompatible with the property B. Obviously, if (1) is an analytical sentence then it means that the object a has not at least one of the properties $A_1, \ldots, A_m$. Therefore it becomes necessary to explain why this object deviates from the ideal type, i.e., why it is deprived of some of the properties $A_1, \ldots, A_m$, e.g., the property A_j. The explanation runs as follows:

(7)
(1) if $C(x)$, then non-$A_j(x)$
(2) $C(a)$
(3) non-$A_j(x)$.

The statements (1) and (2) from schema (7) allow for the explanation of why a given object deviates from the ideal type defined in the statement (1) from the schema (6). For instance, they allow for

. . . a more precise understanding of the non-handicraft components of medieval society (Weber (1), p. 102.)

In this – typical – case the ideal-typical statement does not fulfill any explanatory role – just a heuristic one by stimulating explanation carried out with its help. And since generally this is the case, Weber claims that

The ideal-typical concept will help to develop our skill in imputation in *research*: it is no 'hypothesis' but it offers guidance to the construction of hypotheses. It is not a *description* of reality but it aims to give unambigous means of expression to such a description. (*Ibid.*, p. 90.)

Now it becomes clear that there is no place for concretization or a similar procedure in Weber's model of explanation. In the case (1) of explanation (see the schema (6)) it is superfluous, for the explained property is assigned to objects that fall under the ideal type by definition (they may be called empirical types). In case (2) of explanation (see the schema (7)) the concretization is superfluous, too, for the ideal-typical statement is not employed in that case at all. The Marxian model of explanation will be discussed more carefully later (see Part III of this book) but even on the basis of the materials presented in the first part it is obvious that idealizational laws and their concretizations (or approximations) are necessary elements of the schema of explanation. The idealizational laws reveals the most important factors influencing the investigated phenomenon but their concretizations demonstrate in what way the secondary factors interfere and disturb the pure image. Hence, the law of value T^8 gives us the image of the influence of the most essential factor upon prices and T^7, T^6, etc., add the images of influence of secondary factors (Chapter 1, Section 2). Idealizational laws and their concretizations (approximations) are necessary for the goal of explanation in the Marxian model as well. Instead of this, the Weberian model is based on the assumption that

Substantively, this construct [the ideal type–L.N.] in itself is like a *utopia* which has been arrived at by the analytical accentuation of certain elements of reality. Its relationship to the empirical data consists solely in the fact that where (...) relationships of the type referred to by the abstract construct are discovered or suspected to exist in reality to some extent, we can make the *characteristic* features of this relationship pragmatically *clear* and *understandable* by reference to an *ideal-type* (*Ibid.*, p. 90.)

The Marxian concept of testing is also different than that of Weber. The ideal-typical statements, as analytical ones, are not subjected to any empirical control. It is only the hypotheses about the applicability of these statements that are subjected to empirical testing. As we remember, in Weber's example the hypothesis *h* about applicability of the ideal-typical statement *t* has been rejected, not the statement itself. More generally, from *t* and *h* some observational consequence *b* is deduced; if the consequence turns out to be false, the hypothesis *h* is solely suspected to be false. In the positive result, i.e., when *b* turns out to be true, it is the hypothesis of the applicability of *t* which is confirmed, not the ideal-typical statement *t* itself. The ideal-typical statements are, then, insensitive to empirical control. In Marx's concept this is not the case (see Chapter II, Section 3) – idealizational statements are tested *via* approximation. Experimental procedures, as we shall see (Part III of this book), may be just interpreted as creating conditions that are so close to ideal ones (postulated in the idealizing assumptions of the idealizational statement being tested) as technically possible. If the results of the experiment demonstrate that also the consequent of the idealizational statement is – within the given "interval of tolerance" – satisfied, then the statement is confirmed. If not, it is falsified.

To sum up: the method of idealization differs a great deal in comparison with the ideal-typical method of Weber: as for (1) the range of applicability (2) the logical status of basic statements, (3) the methodological role of those statements.

5. IDEALIZATION AND THE IDEAL-TYPICAL METHOD: PHILOSOPHICAL ASSUMPTIONS

As I have stated, from Marx's point of view the method of idealization presupposes essentialist assumptions about the world: since the actual phenomenon is influenced by principal and secondary factors as well, one has to abstract from secondary factors to establish the idealizational theorem revealing the influence of the principal factors alone (Chapter 2, Section 5). As opposed to this, for Weber the ideal type

. . . is a conceptual construct which is neither historical reality nor even the 'true' reality. (. . .) It has the significance of a purely ideal *limiting* concept with which the real situation or action is *compared* and surveyed for the explication of certain of its significant components. Such concepts are constructs in terms of which we formulate relationships by the application of the category of objective possibility (*ibid.*, p. 93).

Therefore, the ideal – typical concepts

. . . are primarily analytical instruments for the intellectual mastery of empirical data (*ibid.*, p. 106).

Let us try to explicate this difference between Marx's and Weber's conceptions. Let us assume that a given phenomenon F is influenced by two factors H and p; H is the principal factor for F, and p is the secondary factor. Now two investigators try to build the theory explaining phenomena of the type F. According to the Marxian view only one way of abstraction is allowed – that disregarding p. Therefore, the proper idealizational law, according to the Marxian standpoint, should have the form:

(8) if $p = 0$, then $F = f(H)$

Another idealizational statement, in particular the one:

(9) if $H = 0$, then $F = k(p)$

would be incorrect: as we shall see that statement cannot be treated as a relative truth in the sense of the Marxian epistemology. But from the Weberian point of view, accentuating only the ordering function of ideal-types, both statements could be correct: the disregarding of p and the eliminating of H as well. Because their role is not to be present in the internal structure of the phenomenon, but they are intended to be "analytical instruments for the intellectual mastery of empirical data". The statement which would play the function better was a better one from the cognitive point of view, not because it agrees with the internal structure of the phenomenon F in a more adequate way (e.g., demonstrates as the principal factor that one which is in fact the most influential one), but because it plays in a more useful way the role of a convenient instrument. In brief, Marx was an essentialist while Weber was an instrumentalist.

Taking all this what has been said above into account I must conclude that Weber was wrong in claiming that Marx unconsciously employed the Weberian method of ideal types. For Marx employed a completely different method with respect to the methodological characteristics, the range of application and philosophical assumptions.

CHAPTER 4

IDEALIZATION AND POSITIVISM

Another group of methodological doctrines is composed of those which are based on the tradition of empiricism. Let us start with the analysis of positivism. A positivist will be conceived as someone who claims (epistemological) *phenomenalism:* the necessary condition for a sentence to describe phenomena is to remain in a meaning-relation to observational sentences.

It should be added that although the positivist concept of science will be criticised below, I am not underestimating the considerable contribution neopositivism made in the philosophy of science. Applying the precise style of philosophizing neopositivists have made their views clear enough to be effectively criticised, which, in fact, has been done. My criticism has been made possible for the same reason: since I am trying to apply this style of philosophizing which neopositivists had introduced for the first time and apply with so great mastery.

1. THE EXTREME POSITIVIST'S DILEMMA

(a) Extreme Positivism

The thesis of epistemological phenomenalism formulated above contains, as its special case, the thesis of *extreme positivism:* the necessary condition for a sentence to describe phenomena is to be synonymous with a conjunction of observational sentences. Since it is not postulated that the meaning-relation reduces itself to that of synonymity, other, weaker forms of meaning-relation are allowed and, therefore, another form of the thesis of epistemological phenomenalism can be considered, namely that of *moderate positivism.*

Extreme positivism may be found in the earlier Rudolf Carnap's works. His extreme positivism was originally linked to subjectivism conceiving of observational sentences to which all the other statements are to be reduced as sentences about sense data. In *Der Logische Aufbau der Welt* Carnap constructs a "constitutional system" in which sense data are individuals, and the expression "there exists a memorized similarity between *x* and *y*", where *x*, and *y* stand for sensory data, is a primary term (Carnap (1), p. 127). Starting from this introspectional basis the author defines physical, heteropsychological (i.e., referring to other people) and cultural notions. There is a conviction at the basis of this concept that may easily be recognized as a thesis of extreme phenomenalism in its subjectivist version:

Since meaning of every scientific sentence must be reducible to the sentence telling something about something given, then meaning of every notion, no matter what field of knowledge it belongs to, must be reducible to other notions, all the way to the notions of lowest order, which refer to what is directly given. (Carnap, Hahn, and Neurath (1), p. 168.)

The objectivist version – accepted from 1932 on – of the thesis of extreme phenomenalism has been formulated by Carnap in the following way:

Each sentence of every branch of scientific language is synonymous with some sentence of a physical language and may, due to this fact, be translated into a physical language without changing its contents. (Carnap (2), p. 89).

A dilemma arises within extreme positivist views: if the thesis of extreme phenomenalism is true, then why are laws being formulated in empirical sciences? According to the thesis of extreme phenomenalism all laws of empirical sciences – as strictly general sentences, i.e., nonequivalent to any conjunction of observational sentences – are senseless.

This problem was, as is widely known, discussed within the Vienna Circle on a number of occasions. M. Schlick's solution based on extreme phenomenalism is also known:

Natural laws are not sentences, which are either true or false, but they are more like directives, how to make sentences of this kind . . . , they are more like instructions, norms of behavior for an investigator. (Schlick (1), p. 57.)

Another solution consisted in giving up the thesis of extreme phenomenalism, which was done by R. Carnap in the thirties.

It is clear that the thesis of extreme phenomenalism excludes all laws from science, particularly so idealizational ones, at least in so far as laws of science are understood as sentences in a logical sense and not, for instance, as directives. If an extreme positivist talked about idealization as a procedure yielding particular kind of theorems, it would just prove how internally incoherent his concepts are. If even a sentence 'each mammal is a vertebrate' is not logically equivalent to any conjunction of observational sentences, it is all the more so with – for instance – the law of value. But their writings do not give any reason for a charge of incoherence because of this. The problem of idealization does not even occur in them.

Marxian directives were not formulated within extreme positivism and they may not be linked to this view, since idealizational laws – to which all these directives refer – are excluded by the thesis of extreme phenomenalism from the model of science supported by the followers of positivism.

Let us note in passing that comparing the extreme positivist idea of science with the Marxian view of science shows – in spite of the brief and unsystematic character of the latter – how simplified the view of scientific inquiry offered by extreme positivism is, how much the investigator is granted a role of a collector, segregator and 'generalizer' of observations, and how falsified are the real investigator's procedures. I mention this not in order to criticize an orientation practically absent in the present philosophy of science (though still present in the methodological self-consciousness of the specialists of particular empirical disciplines), but in order to point out that the most important charge has not yet been brought against this orientation. And a confrontation of extreme positivism with the presented interpretation of Marxism shows the need for such a charge. The charge runs as follows: the basic reason why the (extreme) positivist model of empirical sciences is an inadequate reconstruction of scientific inquiry is that it does not account for the procedure of idealization which occurs in these sciences, nor is it compatible with the assumption that this procedure is – *de facto* – applied in science. I am not, therefore, criticizing extreme positivism for the reasons its followers gave it up – for allowing "metaphysics" into science and charging some elements of science with "metaphysical" charges etc. but for not allowing the procedure which is basic for empirical sciences in the Marxian model of science.

(b) Operationalism and idealizational notions

A particular kind of thesis of extreme phenomenalism is the *thesis of operationalism:* a necessary condition for a sentence to provide information about reality is to be synonymous with a conjunction of observational sentences describing measuring performances or their results. As P.W. Bridgeman put it,

> a great number of problems we are uncritically posing do not make sense. If a definite problem makes sense, then it must be possible to find an operation through which an answer to it will be found. (Bridgeman (1), p. 28.)

What matters is that a meaningful problem can be expressed with operational terms, i.e., the notions denoting observable acts of measurement or their results. Because

. . . in general we understand by any notion nothing else than a set of operations: a notion is synonymous with a set of operations. (*Ibid.*, p. 5.)

If legitimate notions are the ones which are synonymous with the notions denoting observable measuring performances or their results, then legitimate sentences contain such operational notions, and meaningful problems are the ones which may be answered with the help of such sentences.

We shall not quote popular and convincing criticisms of the operationalist view pointing out, for instance, that it follows from it, that

. . . physics applying the terms which were operationally defined is, in fact, a science of some human behaviour. Its laws speaking seemingly of atoms or electromagnetic waves refer, in fact, to the dependencies between our measuring performances and their results. (Przełecki (1), p. 116.)

I shall, instead, try to show that it is precisely not accounting for idealization which made it impossible for operationalism to create an adequate theory of measurement.

First, an attempt must be made to demonstrate that the standard example of the operationalists – Einstein's definition of simultaneity is not an operational definition since it introduces some concept of an idealizational kind. As we know the postulates of the particular theory of relativity: the principle of relativity and the principle of a constant light velocity hold for inertial systems (Kitajgorodski (1), p. 523.) The concept of inertial system is being defined in such a way that this system

. . . must meet the following condition: a body which is not acted upon by forces, should move with a uniform and straight movement or remain at rest. (*Ibid.*, p. 34).

In order words, the notion of an inertial system is defined in such a way that the law of inertia is emptily fulfilled. The law itself is, of course, an idealizational one since it contains – in its antecedent – the idealizing condition limiting its application to the bodies which are not subjected to any forces. The definition of the concept of an inertial system may therefore be formulated in the following way: u is an inertial system if and only if there occur within it only these bodies which are not subjected to any forces, and move with a steady and straight movement or remain at rest. At least the first condition is an idealizational one. In this case the notion of an inertial system is an *idealizational* one. We call in this way every concept which in a set of meaning postulates characteristic for it contains at least one idealizing assumption; if we deal with a concept explicitly defined (the set of postulates is just an equivalent definition), then it is an idealizational one if one of the conditions establishing the meaning of this notion is an idealizing assumption.

The fact that the notion of inertial system is an idealizational one has far reaching consequences for the problem of the operationalist interpretation of the concept of simultaneity. The concept of simultaneity is refering to the

. . . inertial system alone. (Einstein (1), p. 37.)

In that case the definition of a concept of simultaneity must be preceded with an idealizing assumption limiting the applicability of this definition to inertial systems: if P_1 and P_2 are points in a given inertial system, and they are spatially distant, and in the points P_1, P_2 there occur respectively the events E_1, E_2, then these events are simultaneous if and only if the clocks placed in P_1 and P_2 show the same time of occurence of E_1 and E_2 and they are synchronized with a light signal defined by the following formula: $t_2 = t_1 + (t_3 - t_1)/2$, where t_1 denotes the instant the signal leaves P_1, t_2 denotes the moment it enters P_2 and is reflected, and t_3 the moment when it returns to P_1. Therefore the notion of simultaneity is defined for ideal types of events and may be applied to real ones, i.e., the ones which occur in non-inertial systems, but only by concretization of the above definition; analogically to the law of inertia which may by applied to real bodies by way of concretization. In both cases it is likely approximate concretization which matters. Therefore the more strictly (i.e., with smaller deviations from the above definition) the notion of simultaneity may be applied, the more systems in which the compared events occur approach an inertial one, i.e., the more the sum of forces acting upon the bodies within the system approaches zero.

Let us discuss the question whether the above analysis of the notion of simultaneity is compatible with the operationalist interpretation of Einstein's definition. As Bridgeman points out, prior to Einstein the concept of simultaneity was defined in terms of objective properties of physical objects. But Einstein

. . . criticised the notion of simultaneity, and his criticism was based on the fact, that the observations which make it possible to define two events as the simultaneous ones assume that an observer measured these two events, hence simultaneity is not an absolute property of two events or anything else, but it must also assume the relation of the events to the observer. . . . In this way, analysing what is contained in the notion of simultaneity and considering the observer's act as the essence of the thing, Einstein assumes a new point of view on what concepts of physics should be like, namely – an operationalist one. (Bridgeman (1), p. 8.)

As can be seen it is claimed that the definition of simultaneity provided by Einstein is an operational one, i.e., can be subsumed under this scheme:

(1) If we perform an operation P_1, then an object x has a property Q always and only if we obtain the result P_2 (Przełecki (1), p. 109.)

This view may be criticised by pointing out that Einstein's definition does not have to be formulated in this way, so that the names of actions or results of human actions will occur, and that it is quite unimportant for its physical contents. But something different is being demonstrated here. No matter how the utterance 'operation P_1 on an object y' is understood, it is clear that from the fact that someone, for example an investigator, performed an operation P_1 on an object y it follows that y is a real object, If so, y cannot, of course, meet any idealizing condition, and particularly so, y cannot occur in an inertial system: this would mean that y fulfills non-emptily the law of inertia, which is out of the question. Hence there is no reconciliation between the operationalist interpretation of Einstein's definition and the fact that it is based on the idealizing assumption limiting the applicability of the notion of simultaneity to the events occurring in inertial systems. If Einstein's definition had just these features, i.e., it would be both operational and based on an idealizing assumption, or assumptions, it would just be a definition with a contradictory antecedent, cognitively useless. This conclusion cannot be the result of any methodological interpretation of the definition of simultaneity: hence the operationalist interpretation of this definition must be rejected as inadequate.

Let us add that the notion of simultaneity is not an exception among physical notions. For instance, the temperature is an average kinetic energy of the particles of an ideal gas, i.e., the gas whose particles are material points which do not interact. The notion of temperature is an idealizational one, and its definition takes the following form:

(2) if x is a portion of gas and the dimensions of x are equal to zero and the internal pressure in x is equal to zero, then the

temperature of x is equal to the average kinetic energy of the particles occurring in the portion of gas x.

What is the principle of referring this idealizational notion to the real physical objects is best explained on the example of the unit of luminous intensity – the candela.

The candela is the luminance emitted – in a perpendicular direction – by $(1/6 \times 10^{-5})$ m^2 of a perfectly black body radiating at the temperature of solidification of platinum under 1 atmosphere pressure (Massalski and Studnicki (1), p. 36.)

It is clear that this definition introduces the idealizational notion, since – were it fully formulated – there would occur the condition in the antecedent limiting the applicability of the notion of the candela to perfectly black bodies. The question arises, however, how is the candela used to measure the luminous intensity of real bodies, which are by no means perfectly black ones. As the physicists claim,

There is no perfectly black body in nature, but there are models of surface to be made, whose properties do not differ much from the ones of a perfectly black body. For instance, if a hollow sphere blackened with soot is punctured in one place, then this puncture represents in a very good approximation an ideally black body: a ray which will travel inside through this puncture – will not find its way back. The progress in making perfectly black bodies allowed us to define the unit of luminous intensity – the candela – according to the definition given above. . . . The candela reconstructed in the above way, is laden with a relative error from 0,001 to 0,002. (*Ibid.*, pp. 44–45.)

We shall try to reconstruct the two definitions which are assumed in this passage. The definition of a candela as a unit of luminous intensity for a perfectly black body (*candela*$_1$) is the following one:

(3) if x is a perfectly black body and x is within the temperature of platinum's solidification and x is under the pressure of one physical atmosphere and x radiates in a perpendicular direction, then the candela$_1$ is the luminance emitted by $1/6 \times 10^{-5}$m^2 of the surface of the body x.

This is the definition of the candela as an idealizational notion. In order to measure the luminous intensity of real bodies the definition of candela$_2$ as an 'approximation' of candela$_1$ is introduced. The definition holds for 'almost' black bodies, and the concept 'almost black' has a definite technical meaning – we are speaking of realizable 'models', which, at the time,

. . . present with a good approximation a perfectly black body (*ibid.*, p. 44).

According to this we may formulate the following definition of *candela*$_2$*:*

(4) if x is an almost black body and x remains within the temperature of platinum's solidification, and x is under the pressure of one physical atmosphere, and x radiates in a perpendicular direction, then candela$_2$ is the luminous intensity of the field $(1/6 \times 10^{-5})$ m^2 of the surface of the body x.

The deviation of candela$_2$ from candela$_1$ lies within the admissible, from the nowadays standard accepted limits.

2. THE MODERATE POSITIVIST'S DILEMMA

(a) The Theoretician's Dilemma

As I have mentioned, the meaning-relation must not be one of synonymity, and some of supporters of the thesis of epistemological phenomenalism are inclined to accept different versions of moderate positivism.

Moderate positivists claim that scientific statements include, apart from observational sentences and their generalizations, such semantically uninterpreted formulas whose sole role consists in mediating definite observational sentences (or their inductive generalizations) from other sentences of the same kind. The semantically uninterpreted theorems are described in a different way in various works: we shall call them *formal formulae here.*

Within the moderate positivism there arises the so called theoretician's dilemma formulated by C.G. Hempel: if theoretical sentences (formal formulae) serve their purpose, i.e., they establish links between observable phenomena, then they may be replaced by registering laws, which links directly one kind of observational sentences with the other (particular cases of the antecedent of a given law with particular cases of this law's consequent), hence formal formulae are unnecessary. If formal formulae do not serve their purpose, they are unnecessary as well. Therefore, no matter if the theoretical sentences serve or do not serve their purpose (and they have to do or not) – the theoretical sentence (formal formulae) are unnecessary. Hempel demonstrates that this dilemma arose on the grounds of his own views: let us note that his view coincides with the one of moderate positivism.

Hempel's concept may roughly be presented as follows (Hempel (2), p. 47ff). Let T be a semantically uninterpreted axiomatic system, whose axioms contain exclusively theoretical terms of the vocabulary V_T. Let V_B be a set of observational terms disjunctive with V_T. The interpretational system for a theory T on the basis of observational vocabulary V_B is for Hempel a finite set of sentences I compatible with the theory T, and including as non-logical constants only the expressions from V_T or V_B, and including every term from V_T or V_B in an essential way. Interpreted theory T' is the theory T to whose postulates the new postulates have been joined: namely, the sentence belonging to the interpretational system I. Recalling Craig's theorem the author shows that there exists an axiomatic system T_B which contains only these sentences of the theory T' which comprise only logical constants and observational terms of the set V_B, and is functionally equivalent to the theory T', i.e., "...establishes exactly the same deductive links between the sentences formulated in non-theoretical vocabulary, which the theory T' does" (*ibid.*, p. 50).

As may be noticed, the theoretician's dilemma arises on Hempel's own grounds as it is a subtype of moderate positivism. We may also notice that the very fact that the moderate positivist view does lead to the theoretician's dilemma implies that the view cannot be made compatible with the directive of applying idealizational laws. Stated more simply, it does not allow idealizational sentences into science. Idealizational theories are not interpreted theories in Hempel's sense – their postulates do not (and cannot) include any observational sentence. The assumption that a theorem $t(O)$ which includes an observational term O, belongs to the postulates of an idealizational theory T, leads to contradiction. If T is an idealizational theory, then it follows that the suggested model $M(T)$ of this theory is defined, among others, by the condition that in the domain $M(T)$ all idealizing assumptions of the theory T should be fulfilled. Since idealizing assumptions are not fulfilled in an empirical domain D, then this domain is not the suggested model of the theory T. But the suggested model of the sentence $t(O)$ is just a domain D including real objects. In this case, if we assume that (1) T is an idealizational theory, (2) the sentence $t(O)$ belongs to the postulates of T, then out of the assumption (1) it follows that the suggested model of a theory T is the domain $M(T)$ in which the idealizing assumptions of the theory T are fulfilled, and out of the assumption (2) it follows that the domain $M(T)$ is not the suggested model for the theory T, since it is not the suggested model of one of the postulates of this theory - of the sentence $t(O)$. In this case, no observational sentence may belong to the postulates of an idealizational theory, and if so, then idealizational theories are not subjected to the theoretician's dilemma. They are not interpreted theories in Hempel's sense. The conclusion is that the moderate positivist view does not allow for the directive of applying idealizational laws, and cannot include this kind of directive

Commenting on Hempel's dilemma we may observe that it is based on the assumption that the only aim of a scientific theory is to systematize the experience data, i.e., finding out the relations of consequence between observational sentences. That "terms and general principles of a scientific theory serve their purposes" means for Hempel that "they establish definite relations between observable phenomena" (*ibid.*, p. 41). But then, for someone who thinks scientific theories have different aims, especially for someone who would like to see a scientific theory as a somehow independent source of information about objective regularities, there does not arise a theoretician's dilemma at all, the theoretician's dilemma is formulated with an assumption of an instrumentalist concept of scientific theory; for a realist in treating a cognitive status of scientific theories there is here no dilemma.

(b) Idealizational Notions and CorrespondenceRules

The fact that application of idealizational laws cannot be reconciled with moderate positivism is clearly quite independent of the above reasoning. As has been remarked, within non-observational theorems the moderate positivism allows inductive generalizations and formal formulae. Inductive generalizations (registering laws) cannot be idealizational laws, since the latter cannot be justified by accumulating and generalizing observations (see Chapter 2, Section 4). Registering laws are of the following form 'every *P* is *Q*', where *P* and *Q* are observational terms. Idealizational laws have – roughly speaking – the following form: 'every *F* is *G*', where at least *F* is an idealizational term (for instance, 'perfect gas', 'ideally rigid body', etc.), hence it is a non-observational one, since its designates are not between observable objects (and not between real objects in general). The question arises, whether idealizational laws may, in this case, belong to the next set of non-observational theorems allowed in the (moderate) positivist model of science, namely to the set defined by us as a set of formal formulae. In order to be a formal formula, a theorem should meet some conditions. I shall describe them using the example of R. Carnap's idea from 1956, and then I shall see if idealizational laws meet these conditions.

According to the above mentioned concept of Carnap the language of every empirical science consists of two separately constructed languages: an observational one, L_0, and a theoretical one, L_T. All the extra-logical terms V_0 of the language L_0 are completely interpreted semantically – Carnap assumes that they denote observable properties and relations of the objects which may be observed (Carnap (4), p. 40). Non-logical terms V_T of the language L_T are not semantically interpreted. Their meaning is provided by the postulates of a language L_T alone. The sentences of this language treated as a pure calculus do not say anything about any domain of reality. The relation between these languages is established by the so-called rules of correspondence, which contain both theoretical terms from V_T and observational terms from V_0. These rules are meaning postulates for theoretical terms (*ibid.*, p. 47). Due to the rules of correspondence

> the terms from V_T receive only indirect and incomplete interpretation, since some of them are linked with *C*-rules [the rules of correspondence – L.N.] to observational terms, and the remaining terms from V_T are connected with the former by the postulates from L_T. (*Ibid.*, p. 47.)

As a result, the sentences of the language L_T have some indirect observational meaning: they are not, however – as it has already been noted – directly interpreted in any domain of unobservable objects. In his former work, where above ideas were first put forward, Carnap says that theoretical terms need not be interpreted in a domain composed of unobservable elements, such as electrons, genes, etc. For a theory to function

> . . . it is sufficient to assume that elementary terms [i.e., observational ones – L.N.] have interpretation: an interpretation of other terms is indirectly defined by the formulae of a calculus . . . linking these terms to the elementary ones. (Carnap (3), p. 210.)

A physicist does not need to understand the term 'φ' in the same way he understands observational terms – by knowing a denotation: it is sufficient

> to know how to use the symbol «φ» in a calculus, in order to derive predictions, which may be tested by observation (*ibid.*, p. 211.)

Therefore, Carnap in 1956 assumed the following criterion of empiricality (roughly speaking):

> A theoretical term M has empirical meaning if among the postulates constituting its meaning there are the rules of correspondence which (1) link – directly or by means of other theoretical terms – the term M to some observational one, (2) allow – in conjunction with some sentence containing the term M as the only nonlogical one – to derive deductively some observational sentences (Carnap (4), p. 39ff).

Formal formulae which are permissible in science apart from observational sentences and registrating laws would therefore be constructed of such theoretical terms which have empirical meaning in the sense of the above definition.

Let us note that idealizational laws are not formal formulae in this meaning. They contain idealizational terms which do not have empirical meaning in Carnap's sense. The terms of this kind are being defined as empty in empirical domain, as denoting no real object. So, for instance, the definition of a concept of perfect gas includes two conditions which are idealizing assumptions: that the gas of this kind is composed of particles which are material points and that internal pressure of a portion of this gas is equal to zero. They predetermine the fact that no really existing gas could be a perfect one. What rules of correspondence may therefore be established for a concept of a perfect gas, with what observational terms should be linked to it, if it is clear that no object denoted by observational terms is a perfect gas?

As we see, idealizational laws are not inductive generalizations, nor formal formulae. They do not fit into the model of science presented by moderate positivism. The thesis that idealizational laws and theories are constructed in science is incompatible with the thesis of moderate phenomenalism – idealizational laws are in no meaning relation to observational sentences, not even in a weak one such as postulated by Carnap in 1956.

We are able, therefore, to formulate the *moderate positivist's dilemma:* if the thesis of the moderate phenomenalism is true, then why are idealizational laws formulated in empirical sciences? This dilemma is not even consciously raised by most moderate positivists – they just skip the problems of idealization in their studies in the philosophy of science. Nevertheless, some of the adherents of moderate positivism do undertake the problem of idealization and try to decide it from the point of view of moderate epistemological phenomenalism. Since we have seen that this thesis cannot be reconciled with the directive of applying idealizational laws (and theories), they must understand idealization differently from what I ascribe to Marx – it is not constructing idealizational laws, but some other laws. Looking at such an attempt to reconcile the thesis that idealization occurs in science with the moderate thesis of phenomenalism (like the one C.G. Hempel had undertaken) assures us this guess is right.

(c) Idealizational Statements and Hempel's "Theoretical Idealizations"

By "theoretical idealization" Hempel understands the theorems which are emptily fulfilled, and which are derivable within a given theory as particular cases of more general principles (Hempel (1), p. 81).

"Theoretical idealizations" are, therefore, *ex definitione*, particular cases of the theorems which are described as factual ones. This condition allows us, according to Hempel, to assume that "theortical idealizations" meet the conditions imposed by the positivist philosophy of science as well as all the other types of theorem. Although they are not directly confirmable on the basis of experience, as logical consequences of the theorems which are, they (indirectly) are, also, empirically confirmable.

It is apparent that a set of idealizational laws is mutually exclusive with a set of so defined "theoretical idealizations". "Theoretical idealizations" are what consequents of idealizational laws are in our view. As Hempel himself puts it:

The law regulating behavior of ideal physical systems are deducible from more general theoretical principles, which are well confirmed by empirical evidence: a deduction usually assumes a form of ascribing some extreme values to some parameters of a more general theory. In this way, for instance, laws for perfect gases are deducible from more general principles of a kinetic theory of gases by means of an assumption that volume of gas molecules disappears and there exist no forces of attraction between the molecules – i.e., by accepting that respective parameters are equal to zero (*Ibid.*, p. 80.)

Therefore, if Clapeyron's law takes the following form for us:

(5) if $G(x)$ and $p_w(x) = 0$ and $v_w(x) = 0$, then $p(x)v(x) = NT(x)$

(where G means 'is a portion of gas', p_w is an internal pressure, V_w is the proper volume of the molecules (portions of a gas), p is the external pressure, v is the volume (of a portion of gas), N is a constant, T is the temperature), then Hempel's respective "theoretical idealization" takes the following form:

(6) if $G(x)$, then $p(x)v(x) = NT(x)$.

The law (5), which may also be reformulated by replacing an antecedent with the condition 'x is a portion of a perfect gas' is an analytical sentence for Hempel, and its truthfulness is guaranteed by a meaning of the term 'perfect gas' (*ibid.*, p. 81).

I shall not follow Hempel's reasoning more closely since we have already obtained a conclusion which is important for us: that the ranges of the terms 'idealizational law' and 'theoretical idealization' are mutually exclusive, i.e., that Hempel's attempt does not mean one may join the directive of applying idealizational laws with the views of the moderate positivism, since Hempel does not speak of idealizational laws.

It should be added that introducing the idea at 'theoretical idealization' to the positivist model of science, Hempel does give rise to a contradiction within the methodological theses of moderate positivism. The thesis of confirmation does belong to them: that the sentence *S* logically follows from a conjunction of the sentences $Z_1, Z_2, \ldots, Z_n$ must be understood as a confirmation of a sentence *S* only if every of these sentences has been confirmed itself. But as we have seen from the quotations from Hempel, the 'theoretical idealization' *I* is a logical consequence of some factual theorems *F* and 'extreme assumptions' (such as the assumption occurring in the precedent of the law (5)) *E*. These 'extreme assumptions' are evidently false, however – these extreme values physical magnitudes never assume for any real objects; for no portions of the gas – to recall Hempel's own example – is it true that its particles are material points and that there are no attractions between them. But in this case Hempel introduces some theorems to science which have to follow from some factual theorems (let us assume, from well confirmed ones) and from 'assumptions' which are not confirmed, and, contrary to the rule of confirmation, he claims that these theorems (i.e. 'theoretical idealizations') are also empirically confirmed. Hempel makes a gross inconsequence from the viewpoint of his own methodological assumptions.

3. THE PROBLEM OF EXPLANATION

A general conclusion which I draw from the above analyses is the following one: the directive of applying idealizational laws does not belong to the positivist model of science, nor is it able to, since it is incompatible with the basic assumption of this model – with the thesis of epistemological phenomenalism. This conclusion predetermines the fact that none of the further methodological principles applied by Karl Marx can be transferred to the positivist grounds. All of them, either connected to explanation, testing, rationalization of human actions, etc. (see Part III of the book), are based on the directive of application of idealizational laws showing what purposes these laws should serve in science, in what way they should be controlled, etc. In addition any of these directives to the system of rules composing the positivist model of science would render this directive immaterial – it could not be applied if all the other directives forbade an introduction of idealizational laws into science.

Let us consider this on the example of Hempel's view of explanation. It is usually called a nomologico-deductive model, in order to distinguish the scheme of it:

$$(H) \qquad \frac{\begin{array}{l} L_1, \ldots L_r \\ C_1, \ldots C_s \end{array}}{E}$$

as possessing two characteristic properties: first, laws are necessary ($L_1, \ldots L_r$), second, from these laws and initial conditions ($C_1, \ldots C_s$) the explained sentence (E) logically follows. (Hempel (3)). The Marxian model of explanation assumes that facts are explained starting from the laws (which grasp principal factors alone) and concretizing them (accounting for the successive secondary factors). The question arises if the scheme (H) is not more general than the Marxian one. It is not since the premises L_1, ..., L_r are called laws, and this is not a purely formal characteristic. From the notion of a scientific law accepted by Hempel it follows that laws used in his schema of explanation are of the factual nature. That Hempel does make this assumption can be seen both from his views on scientific laws and from the examples he uses to illustrate his theory of explanation. One of the laws which, in his opinion, is used by physicists to explain the behavior of real objects is the law stating that whenever the temperature of a gas rises, the pressure being constant, its volume increases (Hempel (3), p. 54). In his opinion, that law falls under the general definition of a law of science as a statement of "a uniform connection between different empirical phenomena or between the different aspects of an empirical phenomenon" (*ibid.*, p. 54). Now neither does his example of Gay-Lussac's law fall under Hempel's definition of a law of science nor does it admit an explanation that would fall under schema (H). The first conclusion becomes evident when a full reconstruction of Gay-Lussac's law is made. In handbooks of physics that law is written:

$$(7) \qquad \text{if } p = \text{const, then } v_t = v_0(1 + \gamma t),$$

where p is the pressure of a given amount of a gas, v_t is the volume of that amount of a gas at temperature t, v_0 is the volume of the gas at the temperature 0 and γ is a numerical constant. The physicists claim that the formula (7) does hold for perfect gases only, i.e., such which have the following properties: their particles are material points, and interaction forces between their particles do not work. Hence Gay-Lussac's law would have to be given the following full formulation (see also statement (5) above):

(8) if $G(x)$ and $p_w(x) = 0$ and $v_w(x) = 0$, then if $p(x) = \text{const}$, then $v_t(x) = v_0(x)(1 + \gamma t(x))$.

It can be noted that (8) does not refer in the least to "various emprical phenomena" or to "various aspects of one and the same phenomenon", as it holds for those gases which satisfy the conditions stated in the antecedent, i.e., for perfect (ideal) gases.

The second conclusion follows directly from the above. Here is the special case of the schema (H) applied to the explanation of particular facts:

(H') (1) if $A(x)$, then $B(x)$
(2) $A(a)$
―――――――――
(3) $B(a)$

Due to the definition of the idealizational law, premise (2) in the *explanans* would always, i.e., for any real object a, be false. Hence idealizational laws cannot in fact explain actual phenomena in Hempel's sense of explanation. It is an evidence against that model of explanation which is supported by Hempel, Popper and many other outstanding philosophers of science. This is so because (i) they fail to notice that schema (H') does not hold for the basic group of scientific statements which is formed of idealizational statements, (ii) their model of explanation does not show how explanations are made by reference to idealizational laws, and hence the model does not point to any cognitive reason that would underlie the construction of such laws.

As opposed to this, the Marxian conception of explanation which will be carefully discussed later (see Chapter 7) refers to idealizational laws. As we have seen, the explanation of particular economic facts (e.g., the level of the price of a given commodity) resorts, in Marx's approach, to the idealizational construction starting from the basic idealizational law (e.g., the law of value T^8), through its concretizations (e.g., T^7, T^6, etc). Hence, the basic relationships constituting the Marxian schema of explanation is that of concretization and approximation. The schema may be, thus, reconstructed in this way (see Chapter 1, Section 2):

$$T^8 \dashv T^7 \dashv \ldots \dashv T^2 \rightsquigarrow\!\dashv AT^2 \wedge P \rightarrow E$$

where $\dashv$ is the relation of concretization, $\rightsquigarrow\!\dashv$ is the relation of approximation, $\rightarrow$ means the logical consequence, P are the initial conditions of the factual statement AT^2 and E is the *explanandum*. One may see that this schema of explanation (which will be presented and discussed more fully in Chapter 8) is organised not by the relation of being more general, i.e., not by the relation of subsumption of one factual statement under another one, but by the relation of concretization of the idealizational statement or by the similar relation of approximation of the statement of the kind. Hence the model of explanation contains idealizational laws.

Let us add that in the model (*H*) even Hempel's "theoretical idealizations" cannot occur. As we have seen, they are, *ex definitione*, particular cases of factual statements. If a "theoretical idealization" *I* implied any sentence *E*, then it might be considered an explanation of this sentence only upon the condition that a factual law *F* was formulated which entails logically *I*. But as the statement *E* follows logically from *I*, the latter from a factual statement *F*, then *E* follows logically from the factual statement *F*. This makes clear that the "theoretical idealization" *I* is a superfluous intermediate link in the process of explanation. The aims of science – explanation and prediction, are realized by factual theorems alone.

The conclusion is a short one: within the positivist model of science there is no place for idealization. Hence, the Marxian conception of science interpreted in this book in the idealizational way has to be at variance with the positivist methodology. Now, let us try to state whether the situation will change in the case of the famous opponent of the positivist philosophy of science – Popper's hypotheticism.

CHAPTER 5

IDEALIZATION AND HYPOTHETICISM

The aim of the present chapter is to compare the Marxian idea of science with hypotheticism. In particular, we want to find out whether the Marxian conception of abstraction interpreted as idealization is included or may be included into the conceptual framework of the Popperian theory of science.

1. POPPER'S CONCEPT OF LAWS OF SCIENCE

(a) Idealization and the Problem of Falsification

One may notice that Popper does not discuss the problem of idealization in his investigations of the methodology of empirical sciences. He does not do so, even though validity of his theses would require at least a discussion of the methodological status of laws concerning material points, inertial systems, perfectly rigid bodies, etc. When Popper suggests the criterion of falsifiability as the criterion of demarcation revealing the most important feature of science (Popper (1), p. 41), he does not consider at all how it is possible for those laws to be falsified. The criterion does not point out the relationship between very statements, common in science, referring to perfectly rigid bodies, rational producers, etc, and the basic sentences. Moreover, Popper speaks of falsifiability of definite idealizational sentences failing to see that these are actually idealizational laws. For instance, Kepler's law, "All the planetary orbits are ellipses", is considered by him to be falsifiable (*ibid.*, p. 122). Also he fails to recognize that this is an idealizational law, for as a matter of fact no planet moves along a perfect ellipse, i.e., does not strictly prove Kepler's laws for mutual perturbations of the planets distort the elliptic tracks (Piekara (1), p. 155).

The above remarks should not be considered to mean that we claim idealizational laws to be non-falsifiable. However, I do claim that the problem of falsifiability of idealizational laws does differ from the question of falsifiability of ordinary factual theorems, and had Popper appreciated the role of idealization in the empirical sciences, he would have discussed both these problems separately. For instance, Kepler's law:

(1) if the planets do not influence each other gravitationally, then their orbits are ellipses, and the sun is in one of the focal points of the ellipses,

is falsifiable, but only in an indirect way. It can be subjected to the approximate concretization:

> . . . in the solar system the mass of the sun is considerably greater than the mass of the planets, so that the force with which the run influences the planets is very large compared to the forces with which the planets mutually interact. Therefore the problem of motion of the planets around the sun can be viewed as the problem of two bodies. However, deviations from the motion of a planet caused by other planets can be accounted for by way of approximate calculus. (*Ibid.*, p. 155.)

It is from a respective factual theorem, which is the approximate concretization of the law (1), i.e., it is based on lifting the idealizing assumption that the motion of a planet around the sun results from gravitational interaction between these two bodies alone, that we can deductively derive basic (observational) sentences. This also testifies (indirectly) the law (1).

As can be noticed on the basis of the above discussion, Popper, while discussing general methodology of the empirical sciences, fails to stress idealization. However, in his study *Poverty of Historicism*, Popper claims that one accepts the assumption of rationality in the humanities, and that the assumption states that humans

> would . . . make the optimal use of all available information for the attainment of whatever ends they may have (Popper (2), p. 140).

The assumption is viewed by him as an idealizing one, for he writes that it enables us

> to construct comparatively simple models of their (humans') actions and inter-actions, and to use these models as approximations (*ibid.*, p. 140).

The conclusion can be that the above note is a coincidence, which Popper did not try to make compatible with his general methodological views, or that there is a hidden view of the humanities behind this assumption. The latter case is furthermore supported by the following evidence in Popper's writings: he claims that the possibility of employing a logical method or the method of rational reconstruction in social sciences

> seems to me, indeed, to indicate a considerable difference between the natural and the social sciences – perhaps *the most important difference in their methods*, since the other

important differences, i.e., specific difficulties in conducting experiments . . . and in applying quantitative methods . . . are differences of degree rather than of kind (*ibid.*, p. 141).

Thus the fact that an assumption of rationality is employed in the humanities is supposed to decide a methodological peculiarity of this branch of knowledge, not a substantial one. Had Popper allowed for idealization in natural sciences, the above mentioned peculiarity of the humanities would be substantial, and would correspond to the fact that one does not accept the idealizing assumption of that content in natural sciences (though other kinds of idealizing assumptions are accepted). This approach is supported by the footnote in which Popper continues the above fragment:

Even here it may be said, perhaps, that the use or rational or 'logical' models in the social sciences, or of the 'zero method', has some vague parallel in the natural sciences, especially in thermodynamics and in biology (the construction of mechanical models, and of physiological models of processes and organs) (*ibid.*, p. 141 footnote 2).

Nevertheless, even if the interpretation of the humanities is ascribed to Popper according to which its methodological peculiarity consists of the employment of idealization – conceived of as the assumption of rationality, for Popper does not mention another kind of idealization in the humanities – we have to say that this has not been made compatible with the whole Popperian model of science. For instance the author fails to explain in what sense the assumption of rationality can be falsified. Thus one fails to decide such an elementary – from the viewpoint of the author of *The Logic of Scientific Discovery* – question as the one about the assumption of rationality: does it belong to science as a valid component, or is it a metaphysical insertion?

(b) Dispositional Terms and Idealizational Ones

Likewise, Popper assumes that "all universals are dispositional" (Popper (3), p. 118). i.e., properties denoted by them are revealed either in actual or possible conditions.

This dispositional character of any universal property will become clear if we consider what tests we should undertake if we are in doubt whether or not the property is present in some particular case (Popper (1), p. 425).

Let us presently consider the property denoted by any idealizational concept. If this property is determined by idealizing assumptions, then no actual object has it, and it follows that no such "possible conditions" exist (i.e., conditions which are not ruled out by factual laws we assume at a given time) in which this property could reveal itself. Thus the property in question is not dispositional and – respectively – the idealizational term which denotes it is not a dispositional predicate in the Popperian sense. The conclusion is self-evident – if "all universals are dispositional", and no idealizational term is dispositional, then idealizational terms and laws are inadmissible in the Popperian model of empirical sciences.

It is clear from the above discussion that the status of the assumption of rationality, and the concept of rationality itself, remain clouded in mystery in Popper's views. Both have been introduced against Popper's general methodological concepts – they are contrary to his concept of scientific laws and to his concept of scientific terms. The view of the humanities as based on the assumption of rationality is based on incoherence from the viewpoint of the model of empirical sciences presented by the same author. The general conclusion which follows from this discussion is that Popper's philosophy of science is incompatible with the directive of employing idealizational laws. Having outlined the view of the humanities as based on the assumption of rationality, Popper was committing an inconsistency in terms of the principles of his philosophy.

2. POPPER'S CONCEPT OF TESTING

(a) Popper and Conventionalism

The above discussion decides the following circumstance: Successive directives which comprise the Marxian methodological system become meaningless in the Popperian system. It is not without interest to compare the Marxian directive of testing to the Popperian one, especially in view of the fact that the author of *The Poverty of Historicism* sacrificed much time and space to the problems of testing.

Let us begin with the statement that Popper is often charged with conventionalism. It is supposed to consist in particular of a

> . . . methodological programme of deductivism . . . which does not require . . . a reference to factual perceptive experience in constructing the fundamentals of empirical knowledge, leaving one free to chose with respect to this. It allows basically for a complete independence of the results of observations and experiments in empirical sciences. (Kotarbinska (1), p. 325.)

This charge seems to be unjustified. Popper rejects a thesis which is broadely accepted in contemporary methodology, namely, that the scientists consider – *ceteris paribus* – an experience of a definite perceptional impression to be a sufficient condition for accepting and justifying respective observational (basic) sentence. For

we do not attempt to justify basic sentences by these experiences. Experiences can motivate the decision, and hence an acceptance or a rejection of a statement, but a basic statement cannot be justified by them (Popper (1), p. 105).

The sphere of perceptional experiences is not the "source" of justification for the basic sentences – the other sentences are: scientists consider some basic sentences to be justified as some other sentences are also thought to be justified. For "every basic sentence can be subjected to the tests" and "this procedure has no natural end" (*ibid.*, p. 104). It is not a matter of decision whether the basic sentence is to be considered justified or not: we have to decide in what particular point do we have to break this chain of justifications, likewise it is a matter of decision that those and not other sentences are accepted as the axioms of a deductive system. Thus in Popper's view, incompatibility with the basic sentences is a sufficient condition for falsifying any hypothesis, while compatibility is an indispensable condition of its tentative acceptance. The same holds true for the testing of basic sentences – as "justifying sentences" other basic sentences are employed: what matters is not whether a given basic sentence can be conditionally accepted on the basis of other basic sentences or not, but whether one should continue testing other sentences – referring to still other basic sentences, or not.

(b) Idealizational Laws and Testing

Testing the laws is the case of *modus tollendo tollens*: If (*if t, then p*) *and non-p, then non-t*, where *t* is ". . . the system of statements which may consist of theories and initial conditions" (*ibid.*, p. 76) and *p* is a basic sentence.

By means of this mode of inference we falsify the *whole system* (the theory as well as initial conditions) which was required for the deduction of the statement *p*, i.e. of the falsified statement. Thus it cannot be asserted of any one statement of the system that it is, or is not, specifically upset by the falsification. Only if *p* is independent of some part of the system can we say that this part is not involved in the falsification. (*Ibid.*, p. 76.)

This pattern of testing the statements in empirical sciences is too simple, however; both as far as characteristics of inferring basic sentences from theoretical statements are concerned, and in case of characteristics of the basic sentences themselves. As far as the former is concerned, suffice it to say that the analyses of the Marxian mode of testing idealizational laws conducted elsewhere (see Chapter 2, Section 3; and Part III) demonstrate that the deduction of basic sentences (observational ones) from the laws and initial conditions, as presented by Popper, is just the last stage of testing idealizational laws. Previous stages i.e., concretization of these idealizational laws which results in obtaining the factual laws from which (along with their initial conditions) there follow the basic sentences in question, are neglected in this presentation. However, the latter problem merits closer inspection here.

(c) Idealization and Observation

Let us begin with an example – the measurement of the circumference of the earth accomplished by Erastothenes (Kline (1), p. 14ff.) Erastothenes knew that Alexandria (*A*) was placed north of Syene (*B*) and that the distance between these two cities is 750 km. During the mid summer the Sun's rays fell perpendicularly to the Earth's surface in Syene, which meant, according to him, that the Sun was then situated on the line *OBS* as indicated in Figure 3:

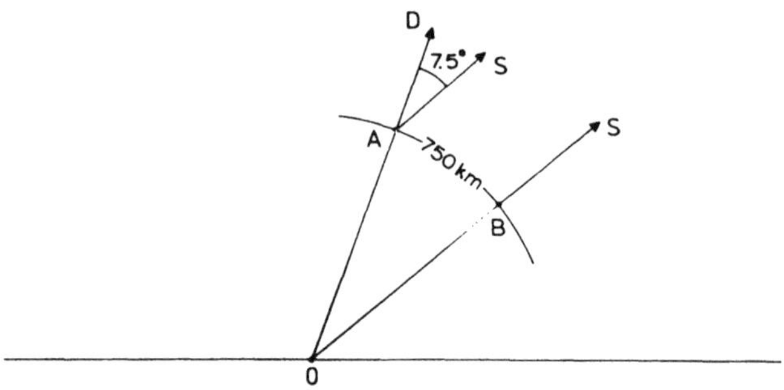

Fig. 3

At the same time the rays fell in Alexandria along the *AS* direction, at an angle *DAS* to the vertical. Erastothenes assumed that the Earth is an ideal sphere, and that as a result of a large distance between the Sun and the Earth, it can be assumed that the straight line *AS* is parallel to the straight line *BS*. Erastothenes measured the angle *DAS* – it was 7.5°. It is clear from geometry, that the angles *DAS* and *AOB* are equal. But *AOB* is 1/48 of a circle. Thus the arc *AB* is 1/48 of the Earth's circumference. As *AB* is 750 km, the circumference of the Earth is about 36,000 km.

It is evident that Erastothenes accepted at least two idealizing assumptions:

p_1: *AS* is parallel to *BS*,
p_2: the Earth is an ideal sphere.

It is only with these idealizing assumptions in mind that he accepted the sentence about the Earth's circumference: $O(Z) = 36{,}000$ km. The approximate nature of this result follows – if the measuring errors in finding the arc *AB* and angle *DAS* are neglected – from the fact that neither p_1 nor p_2 actually hold. Thus, before Erastothenes arrived at the sentence which establishes actual results of measurement, he performed an operation of idealization and concretization. He had primarily accepted the ideal measuring sentence:

(2) If *AS* is parallel to *BS*, and the Earth is an ideal sphere, hen $O(Z) = 36{,}000$ km

where $O(Z)$ is the circumference of the Earth. Erastothenes knew, of course, that the Earth is not an ideal sphere and that in order to obtain a more precise result, he had to introduce a correction z which accounted for the deviation of the Earth's shape from the shape of an ideal sphere. Although this correction is difficult to effectively determine, its theoretical sense is clear. Thus, by removing the assumption p_2 a less abstract measuring sentence is obtained:

(3) if *AS* is parallel to *BC* and the shape of the Earth differs from the shape of a sphere (along the arc *AB* and further on) to the extent of z, then $O(Z) = 36{,}000 \text{ km} \pm z$.

But even sentence (3) is still an idealizational measuring sentence. If p_1 is to be removed, then – as the distance from the Earth to the Sun is much greater than the distance from Alexandria to Syene – an approximate concretization of the sentence (3) can be employed, which makes possible the empirical control of theoretical sentences:

(4) if the straight line *AS* is almost parallel to *BS*, and the shape of the Earth differs (along the arc *AB* and further) from the spheric

shape to the extent z, then $O(Z) \approx 36{,}000$ km $\pm z$.

Since both conditions in the antecedent are realistic ones and *de facto* are met, the factual measuring sentence can be presented as follows:

(5) $O(Z) \approx 36{,}000 \text{ km} \pm z$

It is therefore clear that the procedure of approaching the basic sentence is the following one:

$$(2) \dashv (3) \rightsquigarrow\!\dashv (4) \text{ and } P \rightarrow (5)$$

where $\dashv$ stands for the relation of strict concretization, $\rightsquigarrow\!\dashv$ the approximate one, $\rightarrow$ consequence, while P denotes the initial conditions of the sentence (4).

It is evident, that besides the fact that idealizational laws are concretized in the course of testing, measuring theorems are also concretized.

Thus Popper's scheme of testing – analyzed above – is not only inadequate because it fails to account for peculiarities of the testing of idealizational sentences, but also because it fails to account for the peculiarities of observational procedures, which, at least in the developed empirical sciences, are linked to the idealization and concretization of the process of observation, whereas Popper understands the procedure of observation, and measurement in particular, as a registration of some indications of respective instruments, directed by some questions dictated by a theory which is to be tested (Popper (1), p. 59, footnotes).

(d) Observation and Idealizational Laws

The above remarks refer to the inadequacy of Popper's pattern of testing, according to which testing consists in inferring observational sentences from the laws and their initial conditions. However, it can also be demonstrated that compared with the Marxian directive of testing (Chapters 2, 6 and 10), the directive of the author of *The Logic of Scientific Discovery* turns out to be deficient. It says that if an observational sentence is incompatible with the law, the law (or one of the auxiliary sentences assumed in the course of testing) should be rejected. The Marxian directive of testing allows for some exceptions, which tell us that it is not always the law which has to be discarded in such a situation (especially the idealizational one). If, for instance, somebody decided to concretize idealizational law in an approximate way, and obtained a new factual law, contradicted by one of the observational sentences, the idealizational law in question does not necessarily have to be rejected, if it turns out that this observational sentence contradicts it only *prima facie*, i.e., if by way of the strict concretization we obtain a kind of factual statement which is already not incompatible with the above observational sentence. Both these directives of testing are clearly divergent. Let us decide which gives a better account of an actual control of the laws in empirical sciences.

Here is the description of the procedure of testing the thesis which tells us about the speed of infusion of a perfect liquid through a small hole – from the container with a large and immutable horizontal cross-section:

We cover the hole, fill the container to a definite level, and then, opening the hole, place another container in order to catch the liquid. We measure the time another container is filled and wigh the liquid. We repeat the same routine a couple of times. Results are written down on a table. . . . On the basis of an analysis of errors, we conclude that the speed of inflow can be determined with the precision to 0.01 m/s. Comparing the actual speed of inflow to the theoretical one we expected to obtain on the basis of our hypothesis, we find that the actual speed is always less than the theoretical one, and that the relation of these speeds is, in our case, equal to 0.97 (Wójcicki (1), pp. 56–57).

The hypothesis in question concerning "the speed of inflow of a perfect liquid" is expressed in the formula:

$$w = \sqrt{2gH/\gamma}$$

where w is the speed of the flow, g is the gravitational acceleration, H the weight of the container, and γ the proper weight of the liquid. This equation is valid only for a perfect liquid, that is, for the one which is non-compressible and without viscosity. Hence, x is a portion of the perfect liquid if x is a portion of liquid and the compressibility (c) of x equals zero and the coefficient of viscosity (v) of x equals zero, too. Therefore the hypothesis in question may be interpreted as the following idealizational statement:

(6) $$\text{if } L(x) \text{ and } c(x) = 0 \text{ and } v(x) = 0, \quad \text{then } w(x) = \sqrt{2g(x)H(x)/\gamma(x)}$$

where $L(x)$ means that x is a stream of liquid flowing from a small hole from the container of a large and immutable horizontal cross-section.

Here is the description of the rest of the experiment: from the comparison between the actual and the theoretical speed it is inferred that the general equation should take the following form:

$$w = \alpha\sqrt{2gH/\gamma}$$

where α is the flow coefficient. Looking for reasons for this fact in a more detailed manner, we observe the following phenomenon: the stream flowing out narrows down after leaving the hole. (*Ibid.*, p. 57).

Here is the general conclusion the specialist draws from this case:

This example testifies to the fact that the result of the experiment only partly confirms the hypothesis, for it did not include all the factors which influence the course of the phenomenon we have been able to observe (*Ibid.*, p. 57.)

Let us now try to reconstruct the above example. The statement (6) is tested experimentally. But the experiment concerns, obviously, the imperfect liquid, hence what is compared to the results of the experiment is not (6) but its approximation, i.e., roughly speaking, the statement:

(7) if $L(x)$ and $c(x) \approx 0$ and $v(x) \approx 0$, then $w(x) \approx \sqrt{2g(x)H(x)/\gamma(x)}$.

As the result of the experiment performed on a definite stream of liquid a, the following statement has been accepted:

(8) $L(a)$ and $c(a) \approx 0$ and $v(a) \approx 0$ and $w(a) = b$

while

$$b = \alpha\sqrt{2g(a)H(a)/\gamma(a)} \text{ and } \alpha = 0.97.$$

It follows from this that

$$b \neq \sqrt{2g(a)H(a)/\gamma(a)}$$

Hence the statement (7) is false and should be rejected. It would seem that the sentence (6) should be rejected also. In any case it should be the case had the physicists applied the Popperian directive of testing. However, as we have seen, they do not reject the statement (6), but the only result of the experiment in question is that they decide to subject thesis (6) to concretization. In other words, the physicists claim that the negative result of the experiment is due to the fact that the liquid used in the experiment is not a perfect one, hence the magnitude w was subjected during the experiment to the influence of the factors c and v. At the same time, the experiment showed the correction that should be introduced into the formula occurring in the consequent of (6). Hence, the concretization of (6) is a statement of the form:

(9) if $L(x)$ and $c(x) \neq 0$ and $v(x) \neq 0$, then $w(x) = \alpha\sqrt{2g(x)H(x)/\gamma(x)}$

The statement (8) does not contradict the statement (9); quite the contrary – it confirms the latter. Thus, indirectly, the idealizational statement (6) becomes confirmed. It turns out, therefore, that the result of the experiment only *prima facie* contradicts statement (6): it denies the approximation of (6) but it does not deny a concretization of (6). And, according to the Marxian idea of testing (which will be discussed also in the next chapter), none of the results of observation which only contradict an idealizational statement *prima facie* falsifies the statement. Idealizational statements may be rejected only on the basis of such observations that really contradict them, i.e., deny not only the approximation of a given idealizational statement but also the approximation of each concretization of the statement. As we have seen, physicists apply this rule of testing of statements, because they employ the method of idealization the rule presupposes.

The conclusion that follows from our consideration is simple: hypotheticism, at least in its basic, Popper's version, does not include any grasp of the method of idealization which seems to be, according to the presented interpretation of Marx's methodological ideas, the core of scientific method.

The present result of our comparative analyses is rather strange. Neither so highly formally developed methodological doctrine as logical positivism, nor so philosophically deep concept as the Popperian hypotheticism take into account the procedure which for the first (but unbiassed!) look promised to be one of the most basic and common scientific methods. What is perhaps even more, the method of idealization seems to be the only one universal method applied in all the empirical sciences, starting from physics, through biology, economy to linguistics (think of Chomsky's ideal speaker) or even to jurisprudence (think of the construct of rational legislator) and the science of literature (think of the construct of internal author of a literary work). And this is the space disappear not only the method of induction, but also the method of criticism of hypotheses. The explanation can be only one; there is some philosophical assumption underlying modern methodological doctrine which makes it impossible to see idealization in science. Perhaps, one may guess, the reason is that both positivism and hypotheticism are different versions of modern methodological empiricism according to which experience is always logically prior to theory; in this sense – if they are at variance, experience will supersede theory. This line of reasoning leads to the analyses of the position of the famous opponent of empiricism, T.S. Kuhn, whose philosophy of science is often recognized as the manifestation of 'methodological irrationalism'.

CHAPTER 6

IDEALIZATION AND 'METHODOLOGICAL IRRATIONALISM'

1. THE PRINCIPLE OF EMPIRICISM

The theories of the empirical sciences we have discussed in former chapters maintain that the final criterion of judging whether a given hypothesis can be included into the body of science is experience. Although there are many ways of interpreting experience, methodologists agree at one point – they consider the following principle of empirical testibility to be obligatory in the emprirical sciences:

(1) If it turns out that observed facts contradict an accepted theory, the latter is rejected.

Hypotheticism even based on this principle its concept of science: scientific method, which

> can briefly be described as follows. Faced with a certain problem, the scientists offers, tentatively, some sort of solution – a theory. This theory science accepts only provisionally, if at all; and it is most characteristic of the scientific method that scientists will spare no pains to criticize and test the theory in question. . . . This, of course, is again a variant of the method of trial and error. Theories are put forward tentatively and tried out. If the outcome of a test shows that the theory is erroneous, then it is eliminated; the method of trial and error is essentially a method of elimination. (Popper (3) p. 313.)

Indeed, this principle is of important consequence not only for comperhending the structure of science but also for locating its position in human culture. Hence, the fact that theories are rejected whenever it turns out that the observed world is different from them implies that

(2) scientists have a rational attitude towards accepted statements.

This attitude

> requires that resoluteness with which we speak of our hypotheses and which can be measured by the degree of risk we are ready to take while accepting them is proportional to the degree of their justification. (Ajdukiewicz (2), p. 269.)

This principle of the rational attitude towards accepted statements implies that the scientist's attitude is, in fact, anti-dogmatic. It assumes freedom of thinking:

One has the right and can believe in everything and only in something that is recommended by objective arguments; one is neither obliged nor has to believe in anything which lacks objective argumentation, all the more in anything which is by this argumentation questioned. (*Ibid.*, p. 273).

Therefore,

(3) the society of scientists is radically different from the society of confessors.

The former aims at achieving truth and is ready to give up its statements when they turn out to be false. The latter aims at holding its statements in spite of any criticism. The primary value of the former is to be an innovator and of the latter - a commentator. This uncompromising attitude in striving for truth and readiness for resigning everything which turns out to be true only seemingly is the reason why

(4) there is an unquestionable progress in science.

Since – to quote the most minimalistic criterion of progress:

science is one of the very few human activities – perhaps the only one – in which errors are systematically criticized and fairly often, in time, corrected. This is why we can say that, in science, we often learn from our mistakes, and why we can speak clearly and sensibly about making progress there. (Popper (3), p. 216.)

Of course, everyone realized and does realize that the above image of science and the scientist is never fully carried into effect in reality. However, the deviations from it were treated exactly as deviations – as deviations from typical situations, deviations deserving condemnation and condemned by the scientific group.

2. KUHN'S CRITICISM OF EMPIRICISM

It is clear that the model of science sketched above is based on the principle of empirical testibility (1). No wonder that when, about fifteen years ago, Thomas Kuhn in his book *The Structure of Scientific Revolutions* questioned the above-mentioned statement, using as the base for his argument numerous examples from the history of the sciences, many controversies arose.

The principal role in Kuhn's conception is played by the notion of paradigm. The latter is, roughly speaking, a theory of the domain of facts which at that time became an obligatory pattern of scientific theories. It assumes accepting certain classes of problems as 'scientific' and rejecting other as 'unscientific', using certain investigative techniques, etc. After the pre-paradigmatic period characterized by the existence of numerous competing schools and trends science passes on to the period of maturity. Initially one paradigm is accepted and only problems connected with its development and particularization are taken up. All other problems are considered to be 'unscientific'. Gradually, when anomalies (phenomena incompatible with the paradigm) arise, a crisis in a scientific group takes place. However, the anomalies themselves do not bring about the abandonment of the accepted paradigm. This only happens when a new competitive paradigm originates. The transition from one paradigm to another is called scientific revolution. The domination of the new paradigm begins the next period of 'normal science' which aims exclusively at the development of the paradigm in question.

One of the keynotes of Kuhn's book is the statement that the principle of empirical testibility (1) does not in the least describe the history of science in a sufficient way. Every paradigm from its very origin was opposed by experience so

> If any and every failure to fit were ground for theory rejection, all theories ought to be rejected at all times. (Kuhn (1), p. 145).

Ptolemy himself knew the facts that contradicted his own theory. Up to the modern times, however, it was commonly believed that this theory guarantees the right approach to the problem. It seemed that only minor modifications are necessary to achieve consistency between mathematical predictions and observation. Newton's theory was a similar case. It triumphed, although the facts contradicting it – explained later by the theory of relativity – were well known to everyone. So that

> even a discrepancy unaccountably larger than that experienced in other applications of the theory need not draw any profound response. There are always some discrepancies. Even the most stubborn ones usually respond at last to normal practice. Very often scientists are willing to wait, particularly if there are many problems available in other parts of the field. (*Ibid.*, p. 81.)

The data concerning the history of science presented by Kuhn refuted the principle of empirical testability (1) completely. Therefore, they also refuted the remaining elements of the presented above, in the form of theses (1)–(4), rationalistic model of science. Kuhn's own works can be arguments in favor of the latter statement. For example, questioning statement (2) he maintains that the history of the sciences supplies us with abundant material justifying the following reflexion of Max Planck:

A new scientific truth does not triumph by convincing its opponents and making them see the light, but rather because its opponents eventually die, and a new generation grows up that is familiar with it (*Ibid.*, p. 150).

Consequently, also statement (3) has to be questioned – in fact, scientists are not so much different from religious confessors. They are also believers of their own paradigm for they are also inclined not to pay too much attention to the negative results of experience, to pass over sensible arguments in silence and ignore rivals. The latter are always present – scientists are never exactly of the same opinion concerning the problem of choosing 'the right paradigm' from the alternative theories:

. . . there are always some men who cling to one or another of the older views, and they are simply read out of the profession, which thereafter ignores their work. (*Ibid.*, p. 19).

Finally, Kuhn states that the history of science does not seem to show any permanent progress. There exist only continuous domination and change of paradigm. The whole process, however, takes place without any specified, and fixed for ever, scientific truth whose every subsequent stage of development of scientific knowledge is to be its better and better expression. As the result:

We may, to be more precise, have to relinquish the notion, explicit or implicit, that changes of paradigm carry scientists and those who learn from them closer and closer to the truth (*Ibid.*, p. 169).

It is clear that the implications of Kuhn's views are lethal to the rationalistic model of science. As one of his opponents puts it

For Kuhn, scientific change – from one 'paradigm' to another – is a mystical conversion which is not and cannot be governed by rules of reason and which falls totally within the realm of the (social) psychology of discovery. Scientific change is a kind of religious change. . . . Truth lies in power (Lakatos (1), p. 93).

3. FACTUAL AND IDEALIZATIONAL EMPIRICISM

It is hard to question the facts Kuhn gives to oppose statement (1). Furthermore, anyone who knows scientific circles at least a little can present some more examples. These 'clans' gathered around the master tend to question everything that is stated against him. They stubbornly pass over inconvenient, i.e., not fitting their own conception of the world, ideas in silence and quote only works from a specially acknowledged list of authors. Up to now, however, one tended to regard this situation as a deviation from 'the true scientist'. Kuhn's book seems to point out that a situation of this type is not a deviation but a regular phenomenon.

But the following questions arise: Did Kuhn refute the rationalistic model of science? Should the statement that science to a relatively high degree puts into practice the ideals of several hundred years' old philosophical tradition be included among myths? I think the answer has to be negative. In my opinion both Kuhn and his opponents accept certain similar assumptions. As a result, the latter interpret the principle of empirical testibility (1) in a wrong way and Kuhn misinterprets the collected data concerning the history of science. These tacit assumptions make what may be called a factual conception of science – science (*sc.* empirical) supplies information about regularities in co-existence and sequence of real facts. Although these facts can be unobservable, they can be associated with the observable ones in such a way that when the latter do not take place one is entitled to reject the statement that the former do.

However, as it follows from the first part of the present work, one can find in Marx's writings an outline of another idealizational vision of science. Let us describe it briefly. The scientist distinguishes principal and secondary factors in the investigated reality. About the latter he assumes idealizingly that they do not take place. Therefore, he establishes his basic laws for a simple idealized model in which only principal factors occur. These idealizational laws cannot be applied to reality directly. It can only be done by means of concretization, i.e. by introducing more and more formerly omitted secondary factors, and a proper modification of starting hypotheses. The concretization is carried out up to the point at which the hypotheses are sufficiently precise approximations of the investigated aspect of reality. It can also happen that an approximation of a very abstract idealizational law is false, but an approximation of a slightly less abstract concretization of that law is true.

The principle of empirical testibility (1) is understood narrowly by the advocates of the factual conception of science. A hypothesis concerning usually unobservable aspects of reality is considered. In order to test this hypothesis one has to deduce from it statements that already pertain to observable phenomena and find out whether these theses being particular cases of that statement are true. If not, the starting hypothesis is rejected. Marx interpreted this principle more broadly, so that it did not exclude idealizational statements from the tested hypotheses. He distinguished a category of facts *prima facie* negating the tested hypothesis. A certain fact only *prima facie* negates such a hypothesis when it is incompatible with its approximation and compatible with the approximation of the idealizational hypothesis obtained through concretization of the former. The fact that *actually* negates an idealizational hypothesis is such a fact that contradicts not only the approximation of the tested hypothesis, but also the approximation of every hypothesis obtainable through concretization. In other words, Marx accepted the following principle of empirical testability:

(1′) When scientists assume that the accepted theory is negated by facts only prima facie, they do not reject that theory but try to concretize it. If in spite of that the facts negate also the approximations of the concretized versions of the theory, i.e., it turns out that these facts actually question the starting hypothesis, the scientists reject this theory.

Marx himself used this principle in his own investigative practice. In fact, his *Capital* in its entirety is showing that the facts that seem to negate the basic laws formulated by Marx on a high level of abstraction in the first volume of his work only *prima facie* negate these laws. In his remarks on one of consequences of the law of value he states:

This law clearly contradicts all experience based on appearance. . . . For the solution of this apparent contradiction, many intermediate terms are as yet wanted (Marx (1), Vol. I, p. 307.)

"Intermediate terms" means concretization. Therefore the quoted passage means that although there are "in competition" phenomena that negate the law of value, they no longer negate certain concretizations of the law of value, e.b., so-called law of production prices. Hence, they are only *prima facie* incompatible with the law of value. That is why, according to (1′), the verification of the law is unnecessary. It is clear what Marx meant by writing:

Science consists precisely in demonstrating *how* the law of value asserts itself. So that if one wanted at the very beginning to 'explain' all the phenomena which seemingly contradict that law, one would have to present the science *before* science. It is precisely

Ricardo's mistake that in his first chapter on value he takes *as given* a variety of categories that have not yet been explained in order to prove their conformity with the law of value (Marx (6), p. 196).

It is only the

vulgar economist [who] thinks he has made a great discovery when, in face of the disclosure of intrinsic interconnection, he proudly states that on the surface things look different. In fact, he boasts that he sticks to appearance, and takes it for the ultimate. Why, then, have any science at all? (*Ibid.*, p. 197.)

As we can see, when one accepts the idealizational conception of science, he has to interpret the principle of empirical testability in a slightly different way. One cannot just assume that any deviation from an idealizational law, to be more precise – from an approximation of this law, refutes this law because then every idealizational law and every theory consisting of these idealizational laws would also have to be rejected in advance. It is not difficult to indicate deviations from the law of inertia (when a body is not affected by any force, it is either in regular and rectilinear motion or at rest). The problem is that these observed deviations only at first glance (*prima facie*) contradict this law – Newtonian mechanics can explain them considering less abstract cases, i.e., when forces affecting a body are not necessarily in balance. Instead of principle (1) one has, in fact, to accept principle (1′).

4. REINTERPRETATION OF KUHN'S OBSERVATIONS

Therefore, it seems that a reconsideration of the principle of empirical testability accepted by Kuhn's opponents is necessary. But also interpretation of the facts from the history of science Kuhn refers to while questioning principle (1) has to be changed.

Let us consider the examples Kuhn presents to refute principle (1). It is easy to see that the process of 'development and particularization' of theories is not merely solving successively more specific problems, but a concretization of the starting idealizational theory. For example, Kuhn states:

What to Copernicus was stretching and patching was to them [i.e., to Ptolemy's continuators – L.N.] a natural process of adaptation and extension, much like the process which at an earlier date had been employed to incorporate the motion of the sun into a two-sphere universe designed initially for the earth and stars. Copernicus' predecessors had little doubt that the system would be ultimately made to work (Kuhn (2), p. 75).

This can be expressed in our terminology in the following way. Adherents of Ptolemy's paradigm believed that the facts contradicting their master's theory, like many other facts of this type, will eventually turn out to contradict it only *prima facie*, and that these facts are not excluded from more concretized forms of the theory in question. Copernicus lost his faith and constructed a new theory because he believed that those facts are actually – not only *prima facie* – incompatible with Ptolemy's conception.

In order to reinterpret the abundant data from the history of science given by Kuhn, one has to assume that theories constructed in the sciences are idealizational in character and that developing them is just an attempt at 'approximating' them to actual phenomena. Although Kuhn himself and his opponents do not fully realize it, the above assumption is probably correct. Accepting it, let us reinterpret the following principal from our standpoint Kuhn's statement:

> There are, in principle, only three types of phenomena about which a new theory might be developed. The first consists of phenomena already well explained by existing paradigms, and these seldom provide either motive or point of departure for theory construction. . . . A second class of phenomena consists of those whose nature is indicated by existing paradigms but whose details can be understood only through further theory articulation. These are the phenomena to which scientists direct their research much of the time, but that research aims at the articulation of existing paradigms rather than at the invention of new ones. Only when these attempts at articulation fail do scientists encounter the third type of phenomena, the recognized anomalies whose characteristic feature is their stubborn refusal to be assimilated to existing paradigms. This type alone gives rise to new theories. Paradigms provide all phenomena except anomalies with a theory-determined place in the scientist's field of vision. (Kuhn (1), p. 96.)

If we accept the above-mentioned assumptions, we can easily reinterpret the quoted statement according to principle (1′). Facts of the first type are the ones that can be explained by means of approximation of the starting laws for a given theory – all the more, of course, they are compatible with its further concretizations. Phenomena of the second type are the ones that *prima facie* contradict the theory. They contradict its starting laws, but, as it is proved in the course of the development of paradigm, they do not contradict approximations of more concretized forms of the theory in question. Finally, *anomalies* are the facts that actually contradict the theory; in spite of the adherents' efforts, they stubbornly contradict all concretizations of the theory in the same way they did its starting laws.

As we can see, the discussion between Kuhn and his opponents is based on the same assumption – on the factual conception of science. If we discuss the problem within the idealizational conception, one way expect that statement (1′) is a more adequate formulation of the principle of empiricism than statement (1). Furthermore, while questioning principle (1) Kuhn interprets his counter-arguments according to the factual conception of science. However, one can interpret the data from the history of science according to the idealizational conception of science sketched by Marx in his works. Then the data do not contradict statement (1′) at all – moreover, they even further confirm it. Hence we get the following model of the development of the empirical sciences:

(a) An idealizational theory that already can explain many phenomena is constructed in spite of the fact that at the same time there are many phenomena contradicting it. The latter contradict the approximations of the starting laws of the theory.

(b) The theory in question is concretized. As a result an explanation of many phenomena from the ones that contradict the approximation of the starting laws is obtained. The latter turn out to contradict the theory only *prima facie*.

(c) This results in spreading among the adherents of the paradigm an opinion that all the phenomena that contradict the theory will turn out to be only *prime facie* contradicting it.

(d) Only when more and more phenomena contradicting any concretization of the theory in question are discovered, and, the formerly well-known ones that contradict it cannot be explained in its subsequent concretizations, a crisis takes place. Scientists gradually stop believing that further concretizations of the paradigm will prove that the incompatibilities with experience are only apparent.

(e) The crisis brings about the formulation of a new theory that explains the phenomena that actually contradicted the former one. The new theory becomes also a paradigm, etc. (see Part III).

5. IS IRRATIONALISM NECESSARY?

Let us return to the starting question. Does the above sketched model of the development of science contradict (except for principle (1) which has to be replaced by (1′)) the rationalistic ideals? It does not seem so. It is true that

> the transfer of allegiance from paradigm to paradigm is a conversion experience that cannot be forced. Lifelong resistance, particularly from those whose productive careers have committed them to an older tradition of normal science, is not a violation of scientific standards but an index to the nature of scientific research itself. The source of

resistance is the assurance that the older paradigm will ultimately solve all its problems, that nature can be shoved into the box the paradigm provides. (Kuhn (1), p. 150.)

It turns out, however, that this 'resistance' is thoroughly comprehensible in the light of principle (1′), i.e., in the light of the assumption that paradigms are idealizational theories and their development consists in their concretization. It is clear that scientists do not want to reject their theories as soon as only one of its approximations proves to be erroneous. It is more sensible to assume that probably the theory in question omits certain factors whose interference causes the deviations. Then, finding them out and establishing their range of influence on factors already included in the theory, i.e., the concretization of the theory, is necessary. No one knows in advance whether a given deviation only *prima facie* or actually (i.e., it is an anomaly) does not fit a given theory. Establishing this requires sometimes many years of research. Therefore, from the methodological viewpoint it is more rational to assume that all deviations only *prima facie* contradict the theory and to develop it making best use of all its capabilities. Eventually, this assumptions turn out to be false. However, believing they are correct, scientists solve numerous theoretical and practical problems that could not have been even specified if they had rejected the paradigm thinking its concretization is impossible.

If we reason in the above way, it is clear that none of theses (2)–(4) requires refutation. In my opinion, the scientists can either assume that a deviation does not fit the theory *prima facie*, or think it is an anomaly. None of these alternatives is based on any sensible argument. The deviation does not contradict the idealizational theory that is in force for the idealized world in which secondary factors do not exist; it only signifies that the theory in question cannot be applied to reality through simple approximation. It is still uncertain whether the deviation will contradict further (including secondary factors) concretizations of the theory. This can only be found out after the concretization is carried out. Is not a scientist rational when he concretizes a theory first, instead of assuming it is impossible? Therefore, the scientist accepts statements in the degree of their justification (possible within a given paradigm) according to thesis (2) because the deviations are not counter-examples for an idealizational theory – they only induce one to concretize it. Similarly, it turns out that scientific groups differ from the groups of confessors, in that scientists at least, reject the paradigm when deviations turn out to be counter-examples, when the deviations cannot be explained by any concretization of the paradigm. Finally, it can be stated that subsequent paradigms are, in fact, in some important respects 'more progressive' than their predecessors. Not every new alternative will be accepted as a new paradigm – only the one that satisfies certain conditions, i.e., the one that is a better 'approximation' of reality in the sense of the Marxian concept of relative truth. But this problem will be discussed in Part III.

PART III

AN ATTEMPT AT A SYSTEMATIC CONSTRUCTION OF THE IDEALIZATIONAL CONCEPT OF SCIENCE

CHAPTER 7

ASSUMPTIONS

1. ESSENTIALISM AND ONTOLOGICAL PHENOMENALISM

The conclusion that has been drawn from our previous analyses is that methodology ignores, generally speaking, the method of idealization. The question arises, therefore, why is this method revealed in Marx's methodological remarks, though unsystematic and sometimes unclear as they are? Why does his philosophy pose the problem of idealization? I think that an answer to this is related to the ontological assumptions of the Marxian philosophy.

(a) Essentialist Assumptions of the Method of Idealization

What is characteristic of *Capital* is the incessant stressing of the distinction to be made between essential and adventitious factors, between an inner connection and its manifestation, between what is real and what is apparent. This is not merely a *façon de parler* nor a display of Hegelian phraseology. For Marx it is the issue of carrying out the fundamental duty of science:

. . . it is a work of science to resolve the visible, merely external movement into the true intrinsic movement (Marx (1), Vol. III, p. 307);

. . . all science would be superfluous if the outward appearance and the essence of things directly coincided (*Ibid.*, Vol. III, p. 797), [etc.]

The method which is able to reveal inner connections is called by Marx the method of proceeding from the abstract to the concrete, i.e., it is the method of idealization according to my interpretion. For Marx, as we have ready seen, makes the distinction between two approaches to research:

One of these conceptions fathoms the inner connections, the physiology, so to speak, of the bourgeois system, whereas the other takes the external phenomena of life, as they seem and appear, and merely describes, catalogues, recounts, and arranges them under formal definitions (Marx (5), Part II, p. 165).

The first conception is his own method, i.e, the method of idealization, the second one is the method of induction. The latter is characteristic of the early stage of science, whereas mature theoretical studies are based on the former, and it is then only that science carries out its true task: to reveal the deep structure of phenomena. This is why vulgar economy applying the method of induction

feels particularly at home in the estranged outward appearances of economic relations [which] seem the more self-evident the more their internal relationships are concealed (Marx (1), Vol. III, p. 797).

That connection between the method of idealization and revealing the inner structure of phenomena can also be seen in Marx's analyses of economic issue. Let us consider once more the law of value (see Chapter 1, Section 2). It is based, among others, on the idealising assumption that demand equates supply. Now, as Marx says:

Whenever else this formula [that demand equates supply – L.N.] is resorted to . . . , it serves as a formula to find the fundamental rule . . . which is independent of, and rather determines, competition: notably as a formula for those who are held captive by the practice of competition . . . to arrive at what is again but a superficial idea of the inner connection of economic relations obtaining within competition (Marx (1), Vol. III, pp. 355–356).

Thus the 'formula', i.e., the idealizing assumption, which states that demand equates supply is adopted in order to eliminate the adventitious factor which is the fluctuations of demand and supply, and to concentrate on the essential factor, namely the value of the commodities, that is the time of socially necessary labour required to produce them. Such is in general the role of idealizations: idealizing assumptions are adopted to neutralize certain adventitious factors and thus to enable us to analyse relationships between the given phenomenon and its basic determinants only, that is those which affect most strongly the phenomenon under consideration.

To sum up: for Marx the method of idealization is justified by the fact that reality is essentially differentiated, i.e., the method is the only one that is able to reveal the internal structure of phenomena.

(b) Essential Structure

Let us try to explicate the above-discussed ideas. Let us assume the ordinary set-theoretical apparatus which allows to explicate the notion of a magnitude in usual manner (as a family of classes of abstraction, i.e., of cases of the magnitude) (see Ajdukiewicz (1)). To the apparatus the primitive concept will be added, namely that of the influence one magnitude (factor) has upon another. Given a set U of individual objects, and a set of magnitudes that are constructable on the basis of U, let F stand for any magnitude from that set of magnitudes, and let P_F stand for the set of all those factors magnitudes which have the influence upon F. The set P_F will be termed the *space of essential magnitudes* for F. A magnitude H_i that belongs to the space P_F is an *essential* factor for F. For the sake of simplicity let us assume also that the set P_F is finite for each F:

$$P_F = \{H_1, ..., H_z\}.$$

The ordinary notion of influence is an ordering one, i.e., according to the usual sense of the notion one may say that a given magnitude is more influential for the magnitude under consideration than another one. This is why we introduce the relation of being a more influential factor assuming that it is antisymmetric and transitive in the set P_F. That is, if A is more influential for F than B is, then it is not the case that B is more influential for F than A is. And, if A is more influential for F than B is, and if B is more influential for F than C is, then A is more influential for F than C is. Hence the set P_F is considered to be ordered with respect to the influence of its elements upon F.

In P_F one may also single out factors which are equi-essential for F, namely such that one is not more essential influential for F than the other is. Let those factors which are equi-essential for F form sets

$$D_k, D_{k-1}, ..., D_1, D_0.$$

Factors A, B are in D_i if neither A is more essential (influential) for F than B is, nor B is more essential for F than A is. It follows from the above that the set $\{D_k, ..., D_0\}$ is ordered with respect to the influence of particular factors upon F. That is, all the factors of D_i are more essential for F than those of D_{i+1}, or the reverse holds. Let us adopt the convention according to which the order agrees with the increasing indices, i.e., any factor in D_i is more essential for F than any factor in D_{i-1} is. Those factors which are in D_k are thus the most essential for F of all the factors in P_F. This is why they will be termed *principal factors* for F. Those factors which are in D_{k-1} or . . . or in D_1 or in D_0 will be called *secondary factors* for F. Assume now that the set D_k of principal factors for F consists of $H_1, ..., H_n$.

The set D_{k-1} of factors which are less essential for F than those of D_k are, but more significant for F than the rest of factors in P_F are, consists of the magnitudes $H_{n+1}, \ldots, H_p$. And so on, until we reach the set D_1 consisting of the factors $H_{u+1}, \ldots, H_w$ and the set D_0 consisting of the factors which are essential for F in the least degree: $H_{w+1}, \ldots, H_z$.

Now the hierarchy of factors:

$$
\begin{array}{ll}
(k) & H_1, \ldots, H_n \\
(k-1) & H_1, \ldots, H_n; H_{n+1}, \ldots, H_p \\
 & \ldots\ldots\ldots\ldots\ldots\ldots\ldots\ldots \\
(1) & H_1, \ldots, H_n; H_{n+1}, \ldots, H_p; \ldots ; H_{u+1}, \ldots, H_w \\
(0) & H_1, \ldots, H_n; H_{n+1}, \ldots, H_p; \ldots ; H_{u+1}, \ldots, H_w; H_{w+1}, \ldots, H_z
\end{array}
$$

will be termed the *essential structure* of the magnitude F. The hierarchy will be marked as S_F.

The essential structure S_F contains the set of principal factors and the sets of factors being secondary in different degrees. For the sake of simplicity we adopt the following essential structure as the standard one:

$$
\begin{array}{ll}
(k) & H_1, \ldots, H_n, \\
(k-1) & H_1, \ldots, H_n, p_k, \\
 & \ldots\ldots\ldots\ldots \\
(1) & H_1, \ldots, H_n, p_k, \ldots, p_2, \\
(0) & H_1, \ldots, H_n, p_k, \ldots, p_2, p_1.
\end{array}
$$

The sets of secondary factors

$$D_{k-1}, \ldots, D_1, D_0$$

are, then, assumed to be one-element only (i.e., $p = 1, \ldots, w = 1, z = 1$) and hence marked as p_i ($i = 1, \ldots, k$) in order to distinguish them from the principal factors marked as H_j ($j = 1, \ldots, n$). In our standard form of the essential structure we adopt the following conventions then:

$$
\begin{array}{l}
D_k = \{H_1, \ldots, H_n\} \ (n \geqslant 1) \\
D_{k-1} = \{p_k\} \\
\ldots\ldots\ldots\ldots \\
D_1 = \{p_2\} \\
D_0 = \{p_1\}.
\end{array}
$$

The essential structure S_F is composed of the *levels of significance* (0), (1),

..., $(k - 1)$, (k). Zero level will be called the *surface level*, while the kth level – the *internal* one.

(c) Phenomenon and Essence

Let us consider the fact which consists in the magnitude F assuming certain value n for the object a (in other words, the object a belongs to the case of the magnitude F marked with the number n). The *fact* can thus be treated as a pair (F, a) such that $F(a) = n$, i.e., that a belongs to the case (class of abstraction) of F marked with n. In the extreme degree, F which is the family of classes of abstraction defined on the set of individual objects U, may be a dichotomous magnitude, i.e., it consists of two cases only: Z, $U - Z$; in this situation n may be 1 or 0.

The *complete phenomenal form* of the fact (F, a) will be termed the following system:

$$(H_1, \ldots, H_n, p_k, p_{k-1}, \ldots, p_2, p_1; (F, a))$$

where a belongs to cases of all the magnitudes $H_1, \ldots, H_n, p_k, p_{k-1}, \ldots, p_2, p_1$; that is,

$$\begin{aligned}
&H_1(a) = r_1 \neq 0,\\
&H_2(a) = r_2 \neq 0,\\
&\ldots\ldots\ldots\ldots\ldots\ldots\\
&H_n(a) = r_n \neq 0,\\
&p_k(a) = r_{n+1} \neq 0,\\
&p_{k-1}(a) = r_{n+2} \neq 0,\\
&\ldots\ldots\ldots\ldots\ldots\ldots\\
&p_1(a) = r_{n+k} \neq 0.
\end{aligned}$$

The complete phenomenal form of the fact will be otherwise marked as $(k + 1)$th *phenomenal form* of this fact. For kth *phenomenal form* of the fact (F, a) will be identified with the system

$$(H_1, \ldots, H_n, p_k, p_{k-1}, \ldots, p_2; (F, a))$$

where a belongs to cases of all the magnitudes essential for F with the exception of p_1; that is,

$$\begin{aligned}
&H_1(a) = r_1 \neq 0,\\
&H_2(a) = r_2 \neq 0,
\end{aligned}$$

....

$H_n(a) = r_n \neq 0,$
$p_k(a) = r_{n+1} \neq 0,$
$p_{k-1}(a) = r_{n+2} \neq 0,$

....

$p_2(a) = r_{n+k-1} \neq 0,$
$p_1(a) = r_{n+k} = 0.$

In other words, the least essential factor for F is in the kth *phenomenal form* of the fact (F, a) omitted; etc.

The *first phenomenal* form of the fact (F, a) will be the system

$$(H_1, \ldots, H_n, p_k; (F, a))$$

where a belongs to cases of the principal factors for F and of the most influential secondary factor p_k only; that is,

$H_1(a) = r_1 \neq 0,$
$H_2(a) = r_2 \neq 0.$

....

$H_n(a) = r_n \neq 0,$
$p_k(a) = r_{n+1} \neq 0,$
$p_{k-1}(a) = 0,$

....

$p_2(a) = 0,$
$p_1(a) = 0.$

And the *zero phenomenal form* of the fact (F, a) will be termed the system

$$(H_1, \ldots, H_n; (F, a))$$

where a belongs to cases of the principal factors for F only; that is,

$H_1(a) = r_1 \neq 0,$
$H_2(a) = r_2 \neq 0,$

....

$H_n(a) = r_n \neq 0,$
$p_k(a) = 0,$
$p_{k-1}(a) = 0,$

....

$p_2(a) = 0,$

$$p_1(a) = 0.$$

Now that all the secondary factors for F are eliminated, only the principal factors for F act.

In the case of the complete phenomenal form, a given fact is somehow grasped from the viewpoint of all factors being significant for F, whereas in the case of the zero phenomenal form it is grasped in its 'essential aspect' – only from the viewpoint of the principal factors for F. That is why the former may be identified with the *essence* of the fact (F, a), while the latter can be conceived as the *phenomenon* of the type F.

Notice that the ith phenomenal form of any fact of the type F can be assigned to ith significance level of the essential structure S_F. In other words, what occurs at the ith significance level are ith phenomenal forms of the facts of F type.

(d) Idealization and Discovering the Essence

After the above explications have been made, one may more convicingly present the connection between the method of idealization and revealing the essence which is so strongly stressed in Marx's methodological commentaries. The role of making idealizations, that is, of introducing idealizing assumptions, consists in disregarding in turn secondary factors. As the result of the procedure the phenomenal forms of facts of a given type are revealed one after another; in the extreme case, the zero phenomenal form, or the essence of facts of the type, is discovered. Thus by making the idealizing assumption

$$p_1(x) = 0$$

we disregard the factor p_1. Our considerations then refer to the kth phenomenal form of facts of the type F (in short, of F-facts) including the factor F and the factors $p_2, \ldots, p_k, H_1, \ldots, H_n$. By adopting the next idealizing condition:

$$p_2(x) = 0,$$

we eliminate the next secondary factor p_2. Our considerations then refer to the $(k$-$1)$th phenomenal form of F-facts including the factor F and the factors $p_3, \ldots, p_k, H_1, \ldots, H_n$; and so on, until by making the assumption:

$$p_k(x) = 0,$$

we disregard the most influential of the secondary factors for F. At that stage our considerations refer to the zero phenomenal form, or to the essence, of F-facts which includes only $H_1, \ldots, H_n$, i.e., only the principal factors for F. In this sese the method of idealization makes possible the discovery the essence of facts of a given type, i.e., revealing the influence principal factors have upon the magnitude under investigation. Obviously, the method does not assure by any means that this influence will be in fact revealed. A researcher may be wrong in his recognition of factors which are in fact essential for the magnitude under consideration, for instance, he makes the idealizing condition disregarding the principal factor or he does not eliminate and takes into account a factor which is not essential for F at all, etc. This problem will be discussed later. What is important now is that the method of idealization due to its epistemological nature (disregarding some factors by means of making idealizing conditions) *makes possible* the discovery the essence of facts under consideration. It is another question as to whether possibility is realized or not.

(e) Regularity

There are, in Marxist philosophy, ideas according to which:

A law is a relationship. . . . A relationship of essences, that is, a relation between essences (Lenin (1), p. 127);

A law is an essential phenomenon. Hence, law and essence are homogeneous concepts (of the same order), or, to put it strictly, of the same degree, and express a deeper understanding by man of phenomena, the world, etc. (*ibid.*, p. 126.)

One may perhaps partially explicate these ideas as expressing the following: a *regularity* of F-type is the relationship (dependence) that holds between the magnitude F and the principal factors for it, i.e., it is anyway a function f_k such that:

$$F = f_k(H_1, \ldots, H_n), \text{ if } p_k = 0 \text{ and } p_{k-1} = 0 \text{ and} \ldots \text{and } p_1 = 0.$$

In other words, a regularity is anyway a relation which holds at the inner level of the essential structure. This is what is aimed at by establishing of the idealizational law concerning the magnitude F.

(f) Nomological Structure

Here are the following relationships between the magnitude F and their determinants of the appropriate significance levels of the structure S_F:

$$F = f_{k-1}(H_1, \ldots, H_n, p_k), \text{ if } p_k \neq 0 \text{ and } p_{k-1} = 0$$

and...and $p_2 = 0$ and $p_1 = 0$,

....

$$F = f_1(H_1, \ldots, H_n, p_k, \ldots, p_2), \text{ if } p_k \neq 0 \text{ and...and } p_2 \neq 0 \text{ and } p_1 = 0,$$
$$F = f_0(H_1, \ldots, H_n, p_k, \ldots, p_1), \text{ if } p_k \neq 0 \text{ and ... and } p_2 \neq 0 \text{ and } p_1 \neq 0,$$

where correspondingly:

$$f_{k-1}(H_1, \ldots, H_n, p_k) = g_{k-1}(f_k(H_1, \ldots, H_n), h_{k-1}(p_k)),$$

....

$$f_1(H_1, \ldots, H_n, p_k, \ldots, p_2) = g_1(f_2(H_1, \ldots, H_n, p_k, \ldots p_3), h_1(p_2)),$$
$$f_0(H_1, \ldots, H_n, p_k, \ldots, p_2, p_1) = g_0(f_1(H_1, \ldots, H_n, p_k, \ldots, p_2), h_0(p_1)).$$

The function f_{k-1} is called the first form of manifestation of the regularity with respect to g_{k-1} and h_{k-1}. The function h_{k-1} is the *correctional function* which demonstrates how the secondary factor p_k affects the magnitude F, while the *directional function* g_{k-1} demonstrates the kind of superposition of the regularity f_k and the correctional function g_{k-1}. In an analogous manner one may introduce further forms of manifestation of the regularity, up to its *direct*, or *surface form of manifestation* f_0. The latter reveals the effect upon F of all factors which are essential for that magnitude.

The sequence

$$f_k, f_{k-1}, \ldots, f_1, f_0$$

where f_k is the regularity connected with the factor F and the successive dependencies are its forms of manifestation is called the *nomological sequence* of the factor F. And the following hierarchy of dependencies:

(k) $\quad f_k$,
$(k-1)$ $\quad f_k, (g_{k-1}, h_{k-1})$,

....

(1) $\quad f_k, (g_{k-1}, h_{k-1}), \ldots, (g_1, h_1)$,
(0) $\quad f_k, (g_{k-1}, h_{k-1}), \ldots, (g_1, h_1), (g_0, h_0)$,

will be called the *nomological structure* of the factor F and will be marked as N_F.

(g) Essentialism and Phenomenalism

After designing this conceptual framework one may distinguish between two ontological orientations which are of methodological importance. The *thesis of ontological phenomenalism* says that

(1) the essential structure of every factor is a degenerated one, i.e., it consists of one level only.

A *degenerated essential structure* may be called such a space of essential magnitudes des P_F which is not ordered by the relation of 'being more essential than.' Therefore all the factors that belong to the essential structure of a degenerated nature are equi-essential for the factor F. Thus the degenerated essential structure may be presented in the form:

(0) $H_1, \ldots, H_n, p_k, \ldots, p_1$.

It is therefore senseless to distinguish principal and secondary factors for F if S_F is a degenerate essential structure. Correspondingly, the distinction between the essence and the phenomenon also makes no sense. The distinction between the regularity and its forms of manifestation, the *degenerated nomological structure*, is of the form:

(0) f_0.

That is why the thesis of ontological phenomenalism corresponds to the traditional ideas according to which the distinction between the essence and the phenomenon has no objective meaning but only a subjective one – it expresses our cognitive interest, our needs, and so on.

The thesis of essentialism can be in turn understood as follows:

(2) the essential structure of every factor is a genuine one, i.e., it consists of at least two levels.

A *genuine essential structure* is the one which contains factors differentiated with respect to significance for F (that is, the relation of being more essential than is defined in this case on the space P_F). Therefore the distinctions between the principal and secondary factors, between the essence and the phenomenon between the regularity and its form is in this case meaningful.

2. IDEALIZATIONAL AND FACTUAL CONCEPTIONS OF SCIENCE

The relation between the thesis of essentialism and the necessity of applying the method of idealization is the following. If one believes that the facts are subjected to the differentiated influence of the principal and secondary factors, i.e., if the thesis of essentialism (2) is accepted, then one abstracts from the factors thought to be secondary ones and does take into account factors considered to be principal ones. The latter are accounted for in the starting hypothesis of an idealizational nature, i.e., one being in force under idealizing assumptions eliminating factors treated as secondary for the magnitude under investigation. Since the secondary factors are also essential for the magnitude in question, i.e., they influence it, one does not abstract from them once and for all. Having established a project of the law with idealizing assumptions, an investigator starts to remove these assumptions correcting at the same time the dependence in the consequent of the idealizational hypothesis. In other words, he gradually concretizes it.

Hence, if the thesis of essentialism is assumed, then the method of idealization turns out to be the basic and a quite natural method of cognition. What matters is to 'dissect' by means of hypothetical reasoning the result of observation into 'componential influences', distinguish the strongest influence, and 'put them together' again. If the distinction of the principal and secondary factors is correct and the dependencies (the regularity and its successive forms of manifestation) are recognized properly, then the result is what experience provided us at the beginning, but this time the phenomenon will be theoretically understood. As the Marxian formula has it, from the concrete to the abstract and from the latter to the 'understood concrete' (or 'appropriate concrete'). In this way, we have obtained the answer to the first question, i.e., why the method of idealization can be discovered in methodological reflections of the author of *Capital*. The answer is: because the Marxian philosophy is based on ontological essentialism.

The same reason, being a necessary condition for taking into account the method of idealization, is a sufficient condition for ignoring the method of idealization. That is, the acceptance of the thesis of ontological phenomenalism (1) is a sufficient condition for not having seen in the method of idealization the central method of scientific cognition. As we have seen, ontological phenomenalism was common to all the doctrines that ignore the method of idealization in the Marxian sense. If thesis (1) is accepted, then a fact presents itself as undifferentiated with respect to significance and it calls not for an idealizational reconstruction revealing its internal structure, but it requires a description. The description can be done directly by way of induction or indirectly by way of deduction leading from the general schema of description to particular descriptions of facts – what matters, is that the general theoretical schema are not falsified by the results of observation. The first solution expresses the formula of positivism, the second one – the formula of hypotheticism. What both tendencies have in common, and what they share with the Kuhnian 'anti-empiricism', is the thesis of ontological phenomenalism which generates the factual conception of science. According to it all scientific statements are of a factual nature, explanation employs factual statements only, a theory is a set of factual statements constituted by the relation of consequence, all scientific concepts are introduced with realistic assumptions only, etc. In brief, the ontological phenomenalism generates the factual conception of science.

Therefore, it is surprising that the most commonly employed scientific procedure applying from physics to the science of literature remained unanalysed by very sophisticated methodological doctrines but it was only noticed in modest, unsystematic and sometimes very unclear remarks of the author who treated methodology, as we nowadays view it, as a margin of his occupations. The phenomenalist philosophy assumed by modern trends in philosophy of science made it impossible to recognize a scientific procedure which is based on the contrary ontology. It is a well-known phenomenon that whatever contradicts the accepted theoretical assumptions remains unnoticed or is conceptualized in a wrong way.

3. IDEALIZATIONAL CONCEPTION OF SCIENCE: THE STARTING POINT

So far, we have stated that: (1) Marx employed a definite cognitive procedure which has been described as the method of idealization; (2) this method has not been recognized by modern methodology, for it is based on different philosophical assumptions from the ones they employ. It can, therefore, be supposed that if the method of idealization was made the starting point of construction of a methodological system, then there would be a chance for such a system to represent one of admissible interpretations of the Marxist methodology and to pretend, once it is developed to the sufficient degree, to a systematic explanation of scientific cognition. Our principal task now will be to try to develop the methodological ideas presented in Part I into a systematic methodological system.

(a) The Principal Method

This system will be a kind of idealizational theory of science. Not only in the sense that it will emphasize the role of the method of idealization in empirical sciences but also that it will be built according to this method. The methodological system I want to propose as an interpretation and systematic development of the Marxian ideas will be an idealizational theory itself. If this is so, the problem arises which scientific procedure should be taken into account from the beginning as the principal one and which of the scientific methods may be disregarded in the ideal model of science as being secondary. The follower of the thesis of essentialism has no doubts with respect to the following: if the basic feature of reality is that it is differentiated with respect to significance, then the principal method of cognition of reality has to be the only method which is able to recognize that feature, that is, the method of idealization. All the remaining methods such as the axiomatic method, the method of modelling, of approximation, etc., are to be treated as secondary procedures which must be abstracted from in the initial, the most idealized model. They should be taken into account in further, more also realistic models of scientific cognition. Therefore, the thesis of essentialism also refers to this particular case of scientific cognition, viz. the cognition of cognition; the method of idealization is the essence of cognition, while the other procedure is its appearance which must be accounted for but not sooner than in the course of concretization. Scientific cognition is the only phenomenon which is able to reflect other phenomena.

(b) The Essence of Science

The essence of scientific cognition consists then in the systematic recognition of the essence of investigated facts which is, at the same time, approximately successuful – the indicator of the latter is the growing effectiveness of social practice based on scientific knowledge. The theory of science pretending to develop the ideas of the Marxian philosophy, on the one hand, has to consider the principal means of achieving knowledge about what is essential in reality and, on the other hand, has to be concerned with the conditions of realizing this goal. The latter consists in the fact that scientific cognition is a social phenomenon – knowledge is achieved through a kind of social process. Applying the method of idealization is necessary but not sufficient for successfully approaching absolute truth. Therefore, two ideas are combined in the formula that the essence of science is the approximately effective approach towards the recognition of the essence of facts: (1) Scientific cognition is a reflection of reality, i.e., its internal structure (idealizational laws and their successive concretizations) reflects the internal structure of facts (the essence and its forms); (2) scientific process is a kind of social process.

(c) Scientific Cognition as Reflection of the Reality

According to our assumption that the principal method of science is that of idealization we may treat sequences consisting of the idealizational law and its successive concretizations:

(3) $T^k, T^{k-1}, \ldots, T^1, T^0$

(where the index i shows the amount of idealizing conditions being in force) as the main units of science. In sequence (3) the statement T^i is the concretization of T^{i+1} – therefore T^i is a more realistic presentation than T^{i+1} of what F, the magnitude under investigation, depends on. Now one may notice that statement T^i is supposed to describe the ith level of significance of the essential structure S_F with respect to the repertoire of determinants (essential factors) accounted for and the ith level of the nomological structure N_F with respect to revealed dependencies. It follows from this that the statement T^i is supposed to present the dependence f_i in the nomological sequence of the factor F consisting of the regularity f_k and its successive forms of manifestation:

(4) $f_k, f_{k-1}, \ldots, f_1, f_0.$

It seems to be quite natural to explicate the notion of reflection as the isomorphism of the epistemological sequence (3) with the ontological one (4), i.e., such a situation as when T^i is the concretization of T^{i+1} if and only if f_i is the form of manifestation of f_{i+1} And the isomorphism of the kind is a goal of scientific activity. This does not mean that it is the final effect of the cognitive process: (subjective) goal and (objective) effect belong to different dimensions of science.

(d) Cognitive Process as a Social Process

The formula that cognitive process is a kind of social process became a slogan in Marxism. Therefore, let us refer to the Engelsian characteristics of the notion of social process more carefully:

In one point, however, the history of the development of society proves to be essentially different from that of nature. In nature – in so far as we ignore man's reaction upon nature – there are only blind, unconscious agencies acting upon one another, out of whose interplay the general law comes into operation. Nothing of all that happens . . . happens as a consciously desired aim. In the history of society, on the contrary, the actors are well endowed with consciousness, are men acting with deliberation or passion, working towards definite goals; nothing happens without a conscious purpose, without the intented aim. But this distinction, . . . cannot alter the fact that the course of history is governed by inner general laws. For here, also, on the whole, in spite of the consciously desired aims of all individuals, accident apparently reigns on the surface. That which is willed happens but rarely; in the majority of instances the numerous desired ends cross and conflict with one another, or these ends themselves are from the outset incapable of realisation or the means of attaining them are insufficient. Thus the conflicts of innumerable individual wills and individual actions in the domain of history produce a state of affairs entirely analogous to that prevailing in the realm of unconscious nature. The ends of the actions are intended, but the results which actually follow from these actions are not intended; or when they do seem to correspond to the end intended, they ultimately have consequences quite other than those intended. (Engels (5), p. 365.)

Social process may be, then, characterized by two 'parameters': on the one hand, by the actions undertaken by its participants; on the other, by the objective global effect to which all these actions jointly lead. The former are results of conscious aiming at definite states of affairs, the latter is not intended by participants of the process. Thus in order to explain a social process one has to do the following. First, he has to explain the particular actions referring to the aims they have and the knowledge of agents about means of achieving these aims. Second, he has to reveal regularities which jointly make every participant contribute to the global effect which is usually not actually understood (or sometimes is even undesired).

According to the above remarks, one has to start with an explanation of the aim posed by a typical participant of the cognitive process. The aim is, according to what has been said, discovering the essence of facts; the principal means of achieving the aim is the method of idealization. The global effect of the cognitive process is the amelioration of non-cognitive praxis, primarily the productive one, which, in turn, stimulates further growth of science. It is not the case that a subjective aim of people who come to know the world was (or is) the amelioration of non-cognitive social praxis: the aim was and is to understand the world, i.e., to recognize its internal structure. However, what matters is that an objective consequence of cognitive process was, and is, making this praxis more efficient, which, in turn, stimulates the further growth of science. The problem arises: how it is possible that as a result of actions motivated by a desire to discover the internal structure of facts (with the aid of the method of idealization) the amelioration of non-cognitive praxis follows? This is the problem of the Marxist theory of science.

(e) Theory of Science, Methodology, Theory of Cognitive Practice

The above discussion allows us, in so far as it is an adequate reconstruction of the ideas of the founders of Marxism, to distinguish within the Marxist *theory of science* two domains – the *theory of cognitive action* and the *theory of cognitive practice.* The former, which will be also called *methodology*, investigates the aims posed by scientists and procedures (methods) they employ to achieve those aims. The latter looks for the relationship between the actual results of individual research (guided by aims of the participants of the cognitive process) and the global effects that these individual investigations jointly result in. Now I would like to propose some version of the theory of science in the sense that it will be composed both of a methodological model of scientific activity and of some model of interrelations between the results of the activity and some global social effects. I do not want to propose the theory as *the* Marxist one, but it seems to me that it may be viewed as *an* admissible development of the ideas included in the Marxist tradition, particularly in the works of the founders of Marxism.

CHAPTER 8

THE MARXIAN MODEL OF SCIENTIFIC ACTIVITY (Model I)

At the beginning we shall present the theory of scientific activity based on the supposition that the principal method employed in empirical sciences is that of idealization (Chapters 8–12). Next we shall present an outline of the theory of scientific prectice whose main idea is that the application of the method of idealization within competing scientific orientations leads to the objective, unintended process of approaching the absolute truth which, in turn, results in the amelioration of non-cognitive social praxis.

1. IDEALIZING ASSUMPTIONS OF THE IDEALIZATIONAL THEORY OF SCIENTIFIC ACTIVITY

As it has been said (see Chapter 7, Section 3(a)), the idealizational theory of a typical researcher is, in a double meaning, 'idealizational'. First, as it emphasizes the role of the method of idealization in scientific cognition, and second, as it is itself of an idealizational nature. The latter means that we shall adopt some idealizing conditions that simplify the image of an investigator's activity, i.e., stress some points and disregard other ones. The first model of scientific activity is valid under all the idealizing assumptions, further models will be obtained through removing these assumptions and correcting the models.

Here is the list of the idealizing conditions of model I:

(p_1) The only goal of an investigator is the explanation of facts;

(p_2) all the factors that the investigator thinks to be essential for the magnitude he is interested in are identified by him;

(p_3) the investigator knows (in the sense that he is subjectively convinced) what effect all the factors he considers to be secondary for the magnitude in question have upon this magnitude;

(p_4) he is in a position to secure experimentally conditions in which all the factors treated by him as secondary take on zero value (i.e., the investigator has such technical power that he is able to make perfect experiments);

(p_5) the investigator constructing a theory of the magnitude under consideration does not presuppose any theory of another magnitude.

The counterfactual nature of these conditions is obvious. It is also evident that numerous conditions of the same kind will be implicitly assumed in our conception. Some of them will be discussed later.

All the models will however, consist of methodological hypotheses stating how the ideal researcher, characterized by the assumptions (p_1) – (p_5), carries out his fundamental objectives: how he formulates laws, constructs theories, explains facts and tests his theories.

2. PRE-THEORETICAL AIMS

(a) Categorization of World

The first thing our ideal investigator has to perform, at least tacitly, is the acceptance of a definite repertoire of ontological categories. This is done either by way of employing a ready-made language of a given branch of science or by way of constructing a new language (which is rather rare), since as is known (Suszko (1)) syntactic structure of a language corresponds to ontological categories of reality described in the language.

Having accepted a definite list of ontological categories (for instance, objects, features, relations, events, etc.) our investigator has to categorize the world: he adopts some of these categories as primary and reduces the others to them. For example, the substantialist categorization consists in distinguishing objects (in the extreme, reistic version – only objects, in more moderate versions also other types of objects, e.g., features), while the eventistic categorization consists in distinguishing events as the basic category; the other categories are viewed to be reducible to the basic one.

The categorization of the world (e.g., the theory of relativity presupposes the eventistic categorization of the physical world) is vital for further preceedings of our ideal investigator. It demonstrates what kind of beings our reality consists of, i.e., what beings can assume magnitudes which the investigator is interested in. Having established the substantialist categorization of the world, the investigator is occupied with magnitudes defined on objects (or objects and features); having established the eventistic categorization, he considers only magnitudes defined on events, etc.

Acceptance, then, of a definite categorization of the world results in selecting the following type of domains as the proper domains of the investigator's language:

$$D = (U_1, \ldots, U_g; V_1, \ldots, V_m),$$

where $U_1, \ldots, U_g$ (in some cases $g = 1$) are sets of beings of basic ontological categories, while $V_1, \ldots, V_m$ are magnitudes of a given substantial type (i.e., physical ones or economic ones, etc.) defined on some (or all) of the sets $U_1, \ldots, U_g$.

It is obvious that only our ideal investigator discusses the list of ontological categories and considers the primacy of some of them. Actual investigators usually do not do this, but simply apply the language worked out by generations of predecessors. Only the greatest of them undertake the revision of the acquired language, thus taking part in eternal philosophical controversies.

(b) Ontological Perspective

Having established the type of domain he will be interested in, our ideal investigator has to perform another act involving his philosophical imagination. Aiming at finding out what factors are essential for a given magnitude he is interested in, our researcher considers the type of magnitudes which can influence (be significant for) magnitudes of another type. Thus he starts from general principles of a philosophical nature that will be called the *classifying principles of essential stratification*, i.e., the theses of the following form: factors of the type L are (are not) essential for factors of the type K. On the other hand, the *ordering principles of essential stratification* are of the following form: factors of the type N are more essential for factors of the type K than magnitudes of the type M. The latter refer, of course, only to classes of magnitudes which have been already selected by the classifying principles of stratification.

In order to realize that the theses in question in fact play an important role in empirical sciences, let us look at controversies between 'schools' or 'trends' in science. Consider, for instance, the controversy between Marxist theory and subjectivism in political economy. What matters here is whether one has to start from an isolated individual working with insufficient means towards established goals, or from society as a whole and, what is more, from a society that is structured by class-division. The subjectivists' view was briefly expressed by P. Samuelson:

The best way to formulate simply and convincingly the basic principles of economic theory is to examine economic decisions undertaken by an individual. (Samuelson (1), p. 674.)

Subjectivists start with analyses of the ways an isolated individual acts economically, then consider the results of mutual interrelations of "economic men", in order to draw conclusions about the national economy as a whole. The Marxists hold fast to the principle formulated by the founder of the school:

Individuals producing in a society, and socially determined production – this is naturally the starting point. The isolated single fisher and hunter, who form Smith's and Ricardo's starting point are un-imaginative illusions of 18th century Robinsoniads. (Marx (7), p. 320.)

In accordance with this, the scholarly practice of *Capital* reveals economic relations of the representatives of two basic, antagonistic classes: the industrial capitalist and the worker. This is *Capital*'s starting point that is enriched later on with the representatives of other classes (the merchant, the landowner, etc.) (see Part IV of this book).

One has to realize that subjectivists do not claim that people are socially isolated, that their only goal is to maximize profit, etc. (which is sometimes attributed to them in the Marxist economic literature). They assume this by way of idealization, as Marx presupposes that

according to our assumption – that there is a universal and absolute domination of capitalist production – there are but two classes, that is capitalists and none else then a working one (Marx (1), Vol. II, p. 325).

The controversy in question consists, then, in using the same method, that of idealization, in a different way by both orientations. What a subjectivist abstracts from, when he thinks class relations are secondary features for economic matters, is the heart of the matter for a Marxist, who will not abstract from this. What a Marxist will leave out, regarding the individual's economic strategy as one of the secondary factors, is an important factor for a subjectivist, who will not abstract from it. And so we see clearly what the controversy is about: what is more essential for the economic processes – praxiological strategies of an isolated individual or acts of social classes, i.e., class characteristics of economic agents.

The argument in question may be presented thus in the controversy about validity of the ordering principles of essential stratification:

(1) For economic magnitudes class characteristics of economic agents are more essential than their praxiological strategies,

(2) for economic magnitudes praxiological strategies applied by economic agents are more essential than their class characteristics.

The first is supported by Marxists, the second by subjectivists. Note that for extreme subjectivists even this classifying principle of essential stratification is doubtful:

(3) class characteristics of economic agents are essential for economic magnitudes,

while the radicals among Marxists are inclined to question the truthfulness of the classifying principle:

(4) praxiological characteristics of economic agents are of importance for economic magnitudes.

Our researcher assumes also, tacitly as a rule, certain general theses concerning the kind of the relationships which may hold between magnitudes of the type he is concerned with due to accepted principles of essential stratification. Those theses have the following general form: for every magnitude A in a class K there is a factor B in a class L such that a relationship of a type Q holds between A and B; further, under the principles of essential stratification, only factors of the type L can be essential (in particular, principal) for a magnitude of a type K. Theses of the kind are called *positive principles of ontic connections*; their negations are *negative principles* of the kind.

The instance of an argument about the principle of ontic connections is the controversy between Einstein and the makers of Copenhagen interpretation of quantum mechanics. The principle of determinism defended by Einstein can be interpreted as a principle of ontic connections:

(5) for every physical magnitude there is a set of physical magnitudes such that the former is univocally dependent upon the latter.

If such a set is pointed out and it allows but for a statistical, not univocal determination only, then it must be interpreted in one way only: not all the essential factors for a given magnitude have been found. There are, in consequence, some 'hidden parameters' (i.e., those that are essential for the magnitude under consideration but have not yet been discovered which, disturb the action of basic univocal regularities making them appear as statistical ones. If they are accounted for, this would allow for building a new deterministic theory replacing the quantum mechanics.

The above example refers to so general an ontic connection that it applies to all physical factors. It would be easy, however, to adduce examples of principles of ontic connections that are less general in nature and accordingly hold between magnitudes which are characteristic of some subsets of the set of all objects only. The thesis of psychophysical dualism which excludes the magnitudes that describe states of organism and states of consciousness from the sphere of causal relationship; the principle of teleology (on which the claim of rationality of human nature is based) which establishes goal-oriented relationships between the 'internal world' and the 'external world'; the thesis stating that processes taking place in consciousness are merely 'manifestations' of subconscious impulses; the opinion that sign situations are characteristic of the 'world of culture' only; etc.; all these may be interpreted as principles of ontic connections. They are formulated with greater or smaller precision but this fact cannot veil two things: first, that those principles have as their foundations the principles of essential stratification (and hence assume a differentiation of the world into certain classes of factors which are essential or, in particular, principal for those factors), and, secondly, that those principles state what relationships can hold between classes of factors singled out in such a manner. The system of principles of essential stratification and of the principles of ontic connections based on the former, generated by a given philosophical doctrine is called an *ontological perspective.*

When so conceived, philosophy proves to be of enormous significance for specialized empirical disciplines. This is so because in such disciplines, when we construct a theory, we presuppose, consciously or not, a certain set of principles of both kinds discussed above. They tell us what types of factors are to be considered as essential (and, in particular, principal) for magnitudes of a given type and what types of relationships can be expected to hold between appropriate factors. Those ontological theorems are so general in nature that under normal conditions investigators usually do not realize the fact that they do assume them (hence the prevailing opinion that philosophy is of little significance for specialized disciplines). This is revealed when a theory seems to suggest, contrary to expectations, dependencies of a different kind in a given science than those distinguished by the universally accepted ontological perspective (such as was the case, it seems, with the birth of quantum mechanics). When this happens, the new stage of the development of the science involved is marked by philosophical discussions in which the specialists in the given field themselves become involved.

(c) Image of the Essential Structure

Let us assume that our investigator is interested in a definite magnitude F from the set of factors K. Let the essential structure of this magnitude be the system S_F. Now he aims at a reconstruction of the hierarchy S_F of essential factors for F. He does not, of course, know the structure S_F, but he has various general schemes and patterns for tracing it, namely those expressed in the ontological perspective he presupposes consciously or tacitly. Within the ontological perspective the investigator tries to employ the criteria of significance of an empirical nature (e.g., experimentation, statistical methods, etc. – see Brzeziński (1) and (2)) and to reconstruct the essential structure S_F or, in other words, to establish the image of the structure. He does this establishing *classifying essentialist hypotheses*, i.e., theses of the following form:

(6) the factor A is significant for the magnitude F,

where A belongs to the class of factors distinguished by some accepted classifying principle of essential stratification. Hypotheses of this kind allow to reconstruct the space of essential factors P_F or to construct an *image of the space of essential factors* $I(P_F)$ The image is, then, the set of factors being considered as essential for F by the investigator. Proposing ordering essential hypotheses, i.e., statements of the form:

(7) the factor A is more significant for the magnitude F than is the factor B,

where A, B, fulfill ordering principles of essential stratification, the investigator orders the set $I(P_F)$, which allows him to construct the *image of the essential structure* $I(S_F)$. The image is, then, the hierarchy of factors treated, truly or wrongly, by the researcher as being significant for F in such and such "degree". Here is the schema of the image $I(S_F)$:

(m) $B_1, \ldots, B_r,$
$(m-1)$ $B_1, \ldots, B_r, q_m,$
....

(1) $B_1, \ldots, B_r, q_m, \ldots, q_2,$
(0) $B_1, \ldots, B_r, q_m, \ldots, q_2, q_1,$

It is clear now in what consists the predetermining role of an ontological perspective. Our scientist who formulates classifying essentialist hypotheses of the form (6) will account for these magnitudes alone which are of the types given by classifying principle of significance. Formulating the ordering hypotheses of the form (7) he will account only for the magnitudes of the types defined by the ordering principles of this kind. As a result, the image of the essential structure he establishes will belong to the kind allowed by the accepted ontological perspective. In other words, although only empirical procedures can decide which element of the class of images allowed by the perspective will be chosen, it is only from the latter class that the choice is made. For the researcher omits entirely the possible hierarchies of factors which are at variance with the accepted ontological perspective. For instance, the follower of mechanicism does not even dream of looking for the principal factors for physical magnitudes among non-mechanical properties of bodies, the follower of linguistic structuralism does not think about the search for the determinants of language understanding among the inheritable rules of linguistic competence (as does Chomsky), etc. Each of them works empirically in the area shaped by the ontological perspective he presupposes, not necessarily being aware of this. Philosophy (metaphysics) not only enters science but plays a very important role within it by determining the starting point of theory making. The investigators who assume the same ontological perspective *may*, while those who accept different perspectives *must*, construct different theories. In this sense, philosophical presuppositions establish a framework within which a scientific activity takes place.

(d) The Essential Structure and its Images

If the image $I(S_F)$ of the essential structure contains the same internal level as the structure itself, i.e., if the repertoire of the factors considered to be principal ones for F is identical with the set of principal factors for F, then the image $I(S_F)$ is termed *principally similar* to the structure S_F. If this is not the case, the image is *principally different* from S_F. Let the structure be of the form:

$$\begin{aligned} S_F\colon\ & H \\ & H, p. \end{aligned}$$

Here is a principally similar image of it:

$I_1(S_F)$: H
$\quad H, q$

and here is a principally different image of this structure:

$I_2(S_F)$: p
$\quad p, H.$

3. LAW AND THEORY

(a) *Formulating of Hypotheses*

Having established the image of the essential structure $I(S_F)$, the researcher starts formulating the hypothesis about the magnitude F under the influence of the factor considered to be principal for it. Let the image be of the standard form:

$$
\begin{array}{lll}
I(S_F): & (m) & B_1, \ldots, B_r \\
 & (m-1) & B_1, \ldots, B_r, q_m \\
 & & \ldots\ldots\ldots\ldots\ldots\ldots \\
 & (1) & B_1, \ldots, B_r, q_m, \ldots, q_2 \\
 & (0) & B_1, \ldots, B_r, q_m, \ldots, q_2, q_1.
\end{array}
$$

Then the initial hypothesis is of the form:

(T^m) if $U(x)$ and $q_1(x) = 0$ and $q_2(x) = 0$ and ... and $q_m(x) = 0$, then $F(x) = k_m(B_1(x), \ldots, B_r(x))$,

where k_m is a dependence of the type allowed by the accepted principles of ontic connections. If the image $I(S_F)$ is a genuine (non-degenerated) one, i.e., if it contains at least two levels, then T^m is an idealizational statement. If this is the case, T^m is gradually concretized, that is, the following statements are established:

(T^{m-1}) if $U(x)$ and $q_1(x) = 0$ and $q_2(x) = 0$ and ... and $q_{m-1}(x) = 0$ and $q_m(x) \neq 0$,
then $F(x) = k_{m-1}(B_1(x), \ldots, B_r(x), q_m(x))$
$= s_{m-1}[k_m(B_1(x), \ldots, B_r(x)), t_{m-1}(q_m(x))]$

....

(T^1) if $U(x)$ and $q_1(x) = 0$ and $q_2(x) \neq 0$ and ... and $q_{m-1}(x) \neq 0$ and $q_m(x) \neq 0$,
then $F(x) = k_1(B_1(x), \ldots, B_r(x), q_m(x), \ldots, q_2(x))$
$= s_1[k_2(B_1(x), \ldots, B_r(x), q_m(x), \ldots, q_3(x)), t_1(q_2(x))]$

(T^0) if $U(x)$ and $q_1(x) \neq 0$ and $q_2(x) \neq 0$ and ... and $q_m(x) \neq 0$, then $F(x) = k_0(B_1(x), ..., B_r(x), q_m(x), ..., q_1(x))$
$= s_0[k_1)B_1(x), ..., B_r(x), q_m(x), ..., q_2(x)), t_0(q_1(x))]$,

where the dependencies $k_{m-1}, ..., k_0$ are of the types allowed by the principles of ontic connections presupposed by the investigator. The statements $T^m, T^{m-1}, ..., T^1$ are idealizational in nature, while the statement T^0 which is the final concretization of T^m is a factual one. Our investigator is able to proceed concretization with respect to all the idealizing assumptions obtaining thus the factual statement which is the final concretization of T^m. Due to our simplifying conditions (particularly to (p_3)) he does not need to use any approximative statements about objective dependencies: all his statements have the strict form revealing, correctly or not, the influence of factors being considered essential for the magnitude under investigation.

(b) The Marxian Concept of Law

According to Marx's definition, a

law is an inner and necessary connection between two seeming contradictions. (Marx (1), Vol. III, p. 220.)

Hence a law has three features – it refers to an (1) inner, (2) necessary connection (3) between seemingly contradictory facts. Let us explicate those cheracteristics of a law with reference to a law as the description of an objective regularity.

Due to the definition of the notion of a regularity (see Chapter 7, Section 1(c)) the explication of the first feature of Marx's notion of a law may be easily carried out. The inner nature of the law consists in being referred to the inner dependence, i.e., to the regularity in question. Let us assume that the essential structure of the magnitude F is of the standard form:

$$
\begin{array}{lll}
S_F\colon & (k) & H_1, ..., H_n \\
& (k-1) & H_1, ..., H_n, p_k \\
& & .... \\
& (1) & H_1, ..., H_n, p_k, ..., p_2 \\
& (0) & H_1, ..., H_n, p_k, ..., p_2, p_1
\end{array}
$$

and so is the nomological structure of the magnitude F:

$$
\begin{array}{lll}
N_F\colon & (k) & f_k \\
& (k-1) & f_k, \{g_{k-1}, h_{k-1}\} \\
& & \dots\dots\dots\dots\dots\dots \\
& (1) & f_k, \{g_{k-1}, h_{k-1}\}, \dots, \{g_1, h_1\} \\
& (0) & f_k, \{g_{k-1}, h_{k-1}\}, \dots, \{g_1, h_1\}, \{g_0, h_0\}.
\end{array}
$$

Therefore, the regularity concerning the factor F is the dependence f_k which holds between F and principal factors (and only them), i.e., which holds in the absence of secondary factors:

$$F = f_k(H_1, \dots, H_n), \text{ if } p_1 = 0 \text{ and } \dots \text{ and } p_k = 0.$$

The regularity f_k is, then, the inner connection in this sense that it expresses the relationships between facts of the type F and their essence. In its pure form, the regularity would occur if the influences of secondary factors were eliminated. However, according to the thesis of essentialism (for every factor its essential structure is a genuine one) always some secondary factors occur, so all the regularities manifest themselves as their corresponding forms.

Now, it is seen that under the thesis of essentialism only the statement of the form T^m may be referred to the regularity f_k. The idealizational statements $T^{m-1}, \dots, T^1$ and the factual statement T^0 cannot be referred to the inner connection in question. This is why the inner nature may be ascribed only to the most abstract statement in the sequence $T^m, T^{m-1}, \dots, T^0$. Following this one can notice an interesting feature of a law in the Marxian sense: it is always the most abstract of known statements about a given magnitude. With the exception of extreme cases when the image of an essential structure is degenerated, laws are always idealizational statements. Therefore with the exception mentioned, no laws in the Marxian sense are factual ones. For, under the thesis of essentialism, such statements refer to the surface level of a given essential structure at most, not to the inner level. The latter may be achieved by eliminating the adventitious factors one by one. This process of elimination consists in the adoption of idealizing assumptions. It shows that not all the idealizational statements are laws in the Marxian sense. For instance, T^{m-1} cannot be treated as a law since it refers to the surface level of the essential structure. Only those statements which refer to the deepest level of the essential structure may function as laws, i.e., those which are of the inner character.

We shall now discuss another property of laws, namely we shall explain the meaning of the formulation that a law refers to a relationship between apparently contradictory facts. Let us examine those contexts in which Marx uses the concept of apparent contradiction. In his commentary to one of the "necessary consequences" of the law of value he says:

This law clearly contradicts all experience based on appearance. . . . For the solution of this apparent contradiction, many intermediate terms are as yet wanted (Marx (1), Vol. I, p. 307).

"Intermediate terms" are for Marx a synonym of concretization. Hence the above formulation means that in reality the price P, referred to in the law of value T^8 (see Chapter 1, Section 1) does not equal $l(V)$; and in general terms, the magnitude F, referred to in T^k, is not in fact equal to $f_k(H_1, \ldots, H_n)$. Nevertheless, this disagreement with facts is merely apparent: when law T^8 is concretized, additional factors are taken into account. They reduce the actual deviations between the price P and its successive sequences of determinants. In general terms: the difference between the value of F and $f_k(H_1, \ldots, H_n)$, which occurs in empirical reality, proves "an apparent contradiction" between both factors ($f_k(H_1, \ldots, H_n)$ being that the transformation of magnitudes is a magnitude itself), because the concretization T^{k-1} of the statement T^k makes it possible to reduce difference between the value of F and that of the magnitude $f_{k-1}(H_1, \ldots, H_n, p_k)$, and so does T^{k-2}, etc. It can thus be understood why Marx states that the task of science is to explain how the laws (for instance, the law of value) manifest themselves:

Science consists precisely in demonstrating *how* the law of value manifests itself. So that if one wanted at the very beginning to 'explain' all the phenomena which seemingly contradict that law, one would have to present the science before science (Marx (6), p. 196).

In fact, when formulating the initial idealizational statement we never know the final dependence of the investigated magnitude on all its determinants, but at most we merely point to its dependence on the most essential factors. And if this is so, then observations will always show that the magnitude in question takes on values other than those which follow from that initial idealizational statement. That, however, does not refute this statement if it can be demonstrated that some of the successive concretizations are already in agreement with experience. And it is

the vulgar economist [who] has not the faintest idea that the actual everyday exchange relations cannot be directly identical with the magnitudes of value. . . . And then the vulgar economist thinks he has made a great discovery when, in face of the disclosure of intrinsic interconnection, he proudly states that on the surface things look different.

In fact, he boasts that he sticks to appearance, and takes it for the ultimate. Why, then, have any science at all? (*ibid.*, p. 197.)

To sum up, the considered feature of a law (there is only an apparent contradiction between the determined and the determining factors accounted for in the law) means that (1) the approximation of the law is false, but (2) the said statement can be concretized so that resulting statement is already a sufficient approximation of empirical data (i.e., the approximation of the latter is true). A law is thus a statement that may be substantiated, its disagreement with empirical data being merely apparent.

Consider now the last feature of the law in the Marxian sense, viz. its necessity. These contexts in which Marx uses the term "necessary" as referred to laws show the connection between the concept of essence (internal nature) and that of necessity.

It therefore *follows* of itself from the *nature* of the capitalist process of accumulation . . . that the increased mass of means of production that is to be converted into capital always finds a correspondingly increased, even excessive, exploitable worker population. As the process of production and accumulation advances therefore, the mass of available and appropriate surplus-labour, and hence the absolute mass of profit appropriated by the social capital, *must* grow (Marx (1), Vol. III, p. 214.) [First two italics are mine – L.N.]

Consider now what underlies the concept of law, for instance an economic one, which is

. . . a logical necessity . . . proceeding from the nature of the capitalist mode of production (*ibid.*, p. 209).

If the interpretation of Marx's concept of essence adopted in this book is correct, then the numerous formulations of the above kind may be understood thus: a law is necessary because its consequent follows from the essential body of the knowledge on the basis of which that law is constructed. And the essential body of knowledge is that part of the initial knowledge which is not eliminated by introducing idealizing conditions; that part which is eliminated by it is called the scope of abstraction. The distinction between the scope of abstraction and the essential body of the initial knowledge corresponds in meta-language formulation to the distinction between the factors considered to be secondary and those considered to be principal.

The final conclusion of the above considerations is this: not all general statements of the sequence

$$T^m, T^{m-1}, \ldots, T^1, T^0$$

can be a law in the Marxian sense. It is only the initial statement T^m that is capable of being a law. When $m = 0$, T^m is a factual statement, but aside from this rather exceptional case, it is idealizational in nature. What follows from it is that no concretization of a statement can be a law.

Despite appearances the latter conclusion is confirmed by scientific practice. For instance, the second law of motion of Newton is established for inertial systems and refers to the motion of material points within the systems. Hence a reconstruction of this law presents it as the idealizational theorem:

(8) if $O(x)$ and $S(y)$ and x is placed within y and $D(x) = 0$ and $E(y) = 0$, then $F(x) = m(x)a(x)$

where $O(x)$ is read: x is a physical object; $S(y)$ is read: y is a physical system; D are the dimensions of a body; E is the result of external forces; F is the force applied to a given body; m is the mass; and a is the acceleration.

This is the basic law called sometimes a principle of motion which is distinguished by physicists themselves from 'the second law of motion of Newton for noninertial systems'. The latter accounting for two additional forces of inertia – forces of convection F_c and Coriolis' forces (F_{cor}) is of the form:

(9) if $O(x)$ and $S(y)$ and x is placed within y and $D(x) = 0$ and $E(y) \neq 0$, then $F(x) = m(x)a(x) + F_c(x) + F_{cor}(x)$.

The statement (9) is also called a law but this very fact seems to be unimportant with respect to the substantial distinction between the basic statement (8) and its concretization for non-inertial systems (9). It is a matter of a linguistic convenience whether the statement (8) will be called a law and (9) will not, or whether (8) will be termed a principle and (9) will not, being termed a (usual) law. What matters is that (8) has a priority to (9) in the structure of classical mechanics. The priority is marked here by calling (8) a law.

(c) Simple Idealizational Theory

The sequence of statements

$$T_F\colon T^m, T^{m-1}, \ldots, T^1, T^0$$

where T^m is a law (in the Marxian sense of the term) and T^i is a concretization of T^{i+1} ($i = 1, \ldots, m$) will be called a simple idealizational theory of the factor F. This kind of theory is the only form of scientific theory which our ideal investigator constructs within model 1. He starts from an idealizational statement which he believes to describe a regularity and then he proceeds to concretize that statement in order to discover the successive forms of manifestation of the regularity. He thus obtains the sequence of increasingly realistic statements, which he believes shows the regularity and its successive forms of manifestation up to the surface form. This is so, because, under assumption (p_3), the ideal researcher is in a position to concretize the initial statement with respect to all the idealizing assumptions he himself made in that initial statement. After removing the condition in question the situation will change, that is, our investigator will construct more complicated systems of statements.

(d) Ontology and Methodology

Notice that three basic methodological notions introduced above have their counterparts within ontological notions. Here is the said correspondence:

regularity	law
relation of manifestation of the regularity	relation of concretization of the law
nomological sequence	simple idealizational theory

The correspondence is evident due to definitions of the concepts involved. But it is not an *ad hoc* analogy, since both ontological and methodological concepts satisfy, as I have tried to argue, some intuitions and may be treated as acceptable with respect to their conformity to some philosophical ideas on the one hand, and to scientific practice on the other hand. This correspondence expresses the idea of science as being a reflection of reality. Not in the sense that it happens that science tells the truth, but in the sense that the internal structure of scientific constructs is isomorphic to the structure of basic ontological forms of reality described.

4. EXPERIMENTAL TESTING

(a) The Nature of an Experiment

The problem arises in what way idealizational laws are testified, that is, what is the basis for their acceptance. And so under the assumption (p_4) an investigator can provide a situation in which secondary factors do not work. This satisfies the initial conditions of the idealizational statement, and he can draw the observable consequences of it in order to state whether they really take place in the course of such a perfect experiment. If they do not, then he rejects the idealizational statement he is verifying (on the condition that he accepts the premisses assumed). Such a verification by a perfect experiment is an idealization of actual experimental procedure of verification in science: we cannot obtain an ideal vacuum, hence we produce a 'possibly ideal' vacuum, i.e., the technical vacuum and find out whether idealizational statements are approximately satisfied. Since in model I an investigator does not need to refer to approximation (see the assumption (p_3)) and since he is (on the basis of (p_4)) equipped in technical ability to eliminate all the factors considered to be secondary to the zero degree, the experimental verification occurs within model I in the form of perfect experimentation.

(b) Experience and Theory

It is a quite widespread opinion in the modern philosophy of science that experience depends on accepted theory and that it is a theory that allows for a definite interpretation of the empirical data which otherwise would be blind and lack informative content (Popper (1), Feyerabend (1), Kuhn (1), and others). What I am trying to show here is that (1) the conceptual framework of the idealizational concept of science allows for an explanation of the role of (idealizational) theory in the interpretation of facts, and (2) the latter allows also the avoidance of some false consequences of the view in question (particularly those supported by Kuhn (1))

Let us start with quoting Kuhn's formulations revealing the dependence of empirical data upon the theory accepted by a researcher:

> Seeing constrained fall, the Aristotelian would measure (or at least discuss – the Aristotelian seldom measured) the *weight of the stones*, the *vertical height* to which it had been raised, and the *time required for it to achieve rest*. Together with the *resistance of the medium*, these were the conceptual categories deployed by Aristotelian science when dealing with a falling body. . . . Galileo saw the swinging stone quite differently. . . . He therefore measured only *weight*, *radius*, *angular displacement*, and *time per swing* which were precisely the data that could be interpreted to yield Galileo's law for the pendulum. (Kuhn (1), pp. 122–123.) [My italics – L.N.]

It is seen that the difference between Aristotle and Galileo lies in employing different magnitudes to the description of a given situation. In other words, different magnitudes seem to them to be significant for the phenomenon in question. This leads to the supposition that the way of interpreting empirical data is dependent upon the idealizational theory of a given factor accepted by a researcher.

Let us assume the standard form of the image of the essential structure:

$$\begin{array}{lll} I(S_F): & (m) & B_1, \ldots, B_r \\ & (m-1) & B_1, \ldots, B_r, q_m \\ & & \cdots\cdots\cdots\cdots \\ & (0) & B_1, \ldots, B_r, q_m, \ldots, q_1. \end{array}$$

After Klawiter (1), the following system:

$$(B_1, \ldots, B_r, q_m, \ldots, q_1; (F, a))$$

will be called an interpretation of the fact (F, a). If the repertoire of the principal factors from the structure S_F is recognized entirely (i.e., B_i is identical with H_i and $r = n$), then the interpretation may be called understanding of the fact (F, a). If moreover the whole $I(S_F)$ is identical to S_F, then the interpretation will be termed *complete*. Notice, that the notion of the interpretation of a fact is a methodological counterpart of the ontological notion of the phenomenal form of a fact (see Chapter 7, Section 1(b)). He who interprets the fact aims at the discovery of the phenomenon connected to the fact or, at least, of the essence of the fact. He who understands the fact found its essence.

Concerning this kind of concept of interpretation of facts let us realize three things. First, it becomes evident why there do not exist 'naked facts' in science: the names of magnitudes of the image $I(S_F)$ can be, and usually are, of a non-observational nature. Second, one may understand now how it is possible that the accepted theory participates in interpreting empirical data: the interpretation of them presupposes the structure of the theory if the latter is conceived in the idealizational manner. And third, no new facts are 'created' when interpretation is made, only the actual facts are interpreted in different ways dependent on accepted idealizational theories. The latter conclusion (Klawiter (1), p. 84) follows directly from the above definition. Here are the interpretations

$$(B_1, \ldots, B_r, q_m, \ldots, q_1; (F, a))$$
$$(C_1, \ldots, C_k, t_n, \ldots, t_1; (F, a))$$

that being different are at the same time the interpretations of the same fact (F, a).

The latter remark allows to reject Kuhn's idea that accepting different theories people operate in "different worlds" creating new facts, etc. (Klawiter (1), p.85.)

5. EXPLANATION

(a) The Marxian Concept of Explanation

Marx linked explanation with the following idea: to explain a fact is the same as to show how it is related to its essence. In methodological terms: to explain an observation is the same as to refer it to that law from which the observed fact results. According to this idea explanation follows the schema:

$$T^m \dashv T^{m-1} \dashv T^{m-2} \dashv \ldots \dashv T^1 \dashv T^0 \wedge P \rightarrow E$$

where the sequence

$$T^m, T^{m-1}, \ldots, T^1, T^0$$

is a simple idealizational theory of the factor F, $\dashv$ is the relation of concretization, $\rightarrow$ is the relation of logical entailment, P stands for the initial conditions of the factual statements T^0 and E is the explanandum concerning the value the factor F takes on for a given object. Hence, to explain is the same as to point to the principal factor and then successively introduce to the initial law T^m the secondary factors until the full determination of a given fact is described. If the law T^m recognizes the actual principal factors and their actual influence upon F, then the explanation is *basically correct.* If, in addition to this, all the secondary factors and the corresponding forms of manifestation of the regularity have been shown properly, then the explanation is *completely correct.*

Explanations in *Capital* follow that schema. For instance, when explaining price formation Marx starts from the law of value endowed with numerous idealizing assumptions T^8 and then proceeds to concretize it by introducing successively the statements T^7, T^6 etc. Note that Marx's way of explaining price formation only approximately falls under the schema in question, since the latter is too simplified at the present level of abstraction of our considerations.

From the schema in question follows that the starting point of the process of explanation has to be law in the Marxian sense, i.e., the most abstract statement of the accepted statements concerning a given factor. The law is to be treated as revealing the essence of facts of a given kind.

This condition is obeyed by Marx. It can be seen clearly in the structure of *Capital.* Chapter L of this book, 'Illusion Created by Competition' is concerned with an analysis of economic statements, current in Marx's times, which were empirical generalizations of observations of economic life. There are, for instance, such statements as the ones saying that wages determine prices. Marx does not deny in the least that this generalization is supported by observations, but he claims that generalizations in question have no explanatory value whatever as they are not based on his idealizational theory and even turn out to be false in the light of that theory. He says that in fact

experience shows here that in some sphere of production, at any rate, the average price of a commodity rises because wages have risen, and falls because wages have fallen. But 'experience' does not show that the value of commodities, which is independent of wages, secretely regulates these changes. . . . 'Experience' shows in this case that wages again determine the price. Thus in both cases experience shows that wages determine the prices of commodities. But 'experience' does not show the hidden cause of this interrelation. (Marx (1), Vol. III, p. 846.)

In order to reveal this "hidden cause" of that relationship which is confirmed by observations we have to refer to that complex idealizational construction which is the economic theory expounded in *Capital.* What is more,

a scientific analysis of competition is not possible, before we have a conception of the inner nature of capital, just as the apparent motions of the heavenly bodies are not intelligible to any but him, who is acquainted with their real motions, motions which are not directly perceptible by the senses (Marx (1), Vol. I, p. 316.)

And under the thesis of essentialism (Chapter 7, Section 1(g)), one may have "a conception of inner structure" of something only if he has at his disposal an idealizational theory of facts of a given type, theory whose initial statement is an idealizational law. Now, the difference between the Hempel–Popper model of explanation and the Marxian model may be presented thus: the latter takes into account the role of idealizational statements (in particular, laws) in explanations of facts, while the former does not show any cognitive reason for them (see Chapter 4, Section 3).

(b) The Search for Simplicity

On the basis of the Marxian model of explanation one may better understand why researchers strive to formulate as simple statements as possible. One may say that a universal statement of the standard form assumed here T^i is *simpler* than another statement of the kind T^j if the functional term occurring in the consequent T^i has fewer arguments than the functional term in the consequent of T^j has. This shows that an idealizational law is simpler than any of its concretizations. There is nothing strange in this fact as such a law describes the simplest level of the essential structure which contains the smallest number of factors (the principal factors only). As can be seen, the methodological concept of simplicity corresponds to the ontological one. This observation points to the cognitive reasons of seeking simple theorems, for the discovery of a theorem that is simpler than a given theorem makes us suppose that the new theorem describes factors which are more significant than the factors described by the earlier one, and that the earlier theorem will prove to be merely a concretization of the new one. Hence the search for simplicity is a manifestation of the fact that investigators strive for ever deeper explanations (in the Marxian sense of the term). The search for simplicity is, then, the search for more abstract statements, i.e., those that contain more idealizing conditions than the initial ones.

This concept of simplicity which seems to be quite well adapted to the Marxian notion of explanation may be supported with instance from actual research practice. The following example can prove this. The new version of the boom cycle theory presented by M. Kalecki in 1954 has been regarded as more complicated than the former version of 1935. It has been stated that the changes introduced by Kalecki into his model

> cannot be regarded as an improvement. . . . The last model does not have the simplicity of the earlier version (Allen (2), p. 287).

In 1962 M. Kalecki published a modified, simpler version of his theory of 1954. The main changes include an introduction by the author of some more idealizing assumptions. He wrote:

> During the entire course of reasoning I adhere to the following assumptions: (1) I am concerned with the closed economy, (2) I assume that the working men do not save money, (3) I abstract from the changes of reserves, (4) I disregard the expenses and the state revenues, and (5) I do not take into account the delay in time as far as the food expenses are concerned. The assumptions (3), (4) and (5) may seem rather drastic but they simplify reasoning to a large extent . . . . While discussing the given problems in my book *The Theory of the Economic Dynamics* [1954 – L.N.] I did not introduce these simplifications which complicated the exposition of the theory to a large extent. (Kalecki (1), p. 285.)

The simplification here is formulating idealizational laws equipped with more idealizing conditions. That is, this notion of simplicity corresponds to the notion defined above. Furthermore, scientists are inclined to interpret the notion of simplicity in the essentialist way presented above. For instance, O. Lange proves why Kalecki's model of 1954 does not posses the simplicity that characterized the earlier version of the model:

> It is to be stressed that the new version of Kalecki's model does not have the elegance of the very first version. . . . Without doubt the first model of Kalecki was not fully realistic, but it was characterized by the simplicity and clarity of the presentation of forces acting in the process of an economic development. (Lange (2), p. 285.)

(c) The Problem of the Realistic Interpretation of Idealizational Laws

The creators of Marxism undoubtedly were followers of realism in interpretation of scientific laws. It has been clearly stated in many of Marx's announcements concerning the discovering of the essence. Also Engels opposed the instrumentalist tendencies among physicists.

Since the classics of Marxism accepted the realistic interpretation of scientific theories, the problem of the possibility of the realistic interpretation of idealizational laws becomes very important. Applying the usual semantic notions like that of satisfaction to idealizational statements leads to a clearly unintuitive conclusion that all the idealizational statements are true for actual objects as they are emptily satisfied by them. Hence one needs another concept that would be capable of qualifying idealizational statements. Let us consider the essential structure of the standard form:

$$
\begin{aligned}
S_F\colon\ &(k) && H_1, \ldots, H_n,\\
&(k-1) && H_1, \ldots, H_n, p_k,\\
& && \cdots\cdots\cdots\cdots\\
&(0) && H_1, \ldots, H_n, p_k, \ldots, p_1,
\end{aligned}
$$

and assume that at the internal level of the structure there occurs the regularity f_k. A statement of the form

(10) if $\alpha(x)$, then $F(x) = f_k(H_1(x), \ldots, H_n(x))$

will be called *representing the regularity* f_k. As it is seen, conditions $\alpha(x)$ occurring in its antecedent may be of an arbitrary form – they may contain idealizing assumptions eliminating factors $p_k, \ldots, p_1$ or not. What matters is the form of the consequent. In a similar manner the notion of representation of a form of a regularity can be defined. For instance, a statement of the form:

(11) if $\beta(x)$ and $p_k(x) \neq 0$, then $F(x) = f_{k-1}(H_1(x), \ldots, H_n(x), p_k(x))$

is said to be the *representation of the dependence* f_{k-1}. The antecedent of this statement may be quite arbitrary except for the removed condition eliminating the influence of p_k. It is seen that the statement representing the regularity presupposes the image of the essential structure which is basically similar to the structure itself, while the statement representing the surface form of manifestation of the regularity presupposes the image being identical to the structure itself.

Note that the notion of representation of regularities, even though it assumes that a given image $I(S_F)$ is basically similar to S_F, does not assume that a statement which represents a regularity must be satisfied non-vacuously. In other words, a statement which represents a regularity may be an idealizational statement. Since investigators strive for the establishment of regularities and their forms. Such a regularity holds between the factor F and its principal factors $H_1, \ldots, H_n$ (and those factors only). But both F and $H_1, \ldots, H_n$ do occur in fact, and since this is so, then there occur in fact the dependence f_k such that F is an f_k-transformation of the principal factors $H_1, \ldots, H_n$. And this is precisely what is stated in the consequent of the statement:

(12) if $U(x)$ and $p_1(x) = 0$ and...and $p_k(x) = 0$,
then $F(x) = f_k(H_1(x), \ldots, H_n(x))$.

It is in fact so in empirical reality that F is affected by $H_1, \ldots, H_n$ in f_k-way and this is simply what is stated by (12). But it is also so in empirical reality that F is additionally affected by secondary factors $p_k, \ldots, p_1$. This is why the results of observations or measurement of F always lead us to the conclusion that there is no f_k-relationship between F and all its determinants, i.e., $H_1, \ldots, H_n, p_k, \ldots, p_1$. But formula (12) does not state in the least that F is f_k-dependent on *all* the factors which are essential for F. It merely states that F is f_k-dependent on some of its determinants, namely on $H_1, \ldots, H_n$. The results of observations do not refute that if the observations are proceeded in natural conditions. Since observation is always concerned with a complex situation, when not only $H_1, \ldots, H_n$ act but also all the secondary factors $p_k, \ldots, p_1$, we can find out whether, in the world we live in, F is f_k-dependent on $H_1, \ldots, H_n$ only in an indirect way – by gradually introducing secondary factors and by making sure whether in the last resort we can in this way arrive at an agreement with observation. It is only in the idealized model I when the ideal researcher is able to find out this in an direct way making the perfect experiment consisting in omitting all the factors considered to be secondary and stating whether the influence of the factors believed to be principal ones is in fact the same as the tested statement says.

Thus, idealizational statements may be referred to facts in the sense that their consequents may represent actual regularities. It is another question as to whether we are in a position to state this but our impossibility of being sure about it (in view of the impossibility of carrying out a perfect experiment) does not authorize us to project that impossibility onto reality. Yet it may be so that a given idealizational law represents a regularity, even though we cannot find it out directly.

(d) Theoretical Aims

From the hierarchy of pre-theoretical aims: accepting some categorization of the world – assuming a definite ontological perspective – establishing an image of the essential structure one has to distinguish the hierarchy of theoretical aims. It consists of the following sequence of goals of a typical researcher: formulating the hypothesis and its testing – establishing a law – theory construction – explanation.

(e) The 'Context of Discovery' and the 'Context of Justification'

The dichotomy between these two 'contexts' cannot be supported within the idealizational theory of science. The chain of pre-theoretical aims could be considered a counterpart of the 'context of discovery', while the chain of theoretical aims – of the 'context of justification'. However, these chains are not independent of each other. First of all, establishing an image of the essential structure is undoubtedly subordinated to formulating an initial idealizational hypothesis: the latter presupposes the former since the hypothesis accounts for the principal factors in its consequent and abstracts from the secondary factors in its antecedent. Establishment of the image does not determine univocally the form of hypothesis, hence stating the hypothesis is undoubtedly a creative act – both the image and possible empirical data do not show the dependence but at most the type of dependencies. But it is by no means an arbitrary act, as supposed by hypotheticism: only those hypotheses are proposed which are allowed by the image (and indirectly by the ontological perspective and the assumed categorization of the world) and possible empirical data – if the latter are available in a given case at all. Secondly, the reverse dependence of the two kinds of aims also holds – the success of a hypothesis confirms adequacy of the accepted image of the essential structure, its failure makes us reject it. Hence, two contexts are connected to each other and there cannot be any reason for methodology to limit its interest to the 'context of justification'.

6. IDEALIZATION AND TRUTH

(a) Science and Caricature

The comparison between science and caricature is, according to common-sense prejudices, unfair to science and senseless. Science, according to the same belief, has to present the world as it is, while caricature distorts the world in order to emphasize something, not in order to arrive at the truth. The idealizational grasp of science leads, however, to a quite different image of scientific activity – science consists of the same procedure we find in caricature: of the distortion of reality.

Let us see what a cartoonist does: he leaves out some details of the person presented, thus stressing what he considers important. That is, he employs the method of exaggeration: he does not present everything but distorts a person or a situation by neglecting some features he thinks minor ones. Science, as we have seen, in fact does the same. When a physicist constructs the concept of a material point, he does not present physical objects but distorts them – he assumes that they have zero dimensions and focuses on other properties of these bodies (eg., mass) which he considers more essential for physical magnitudes he investigates. In short: science consists in the same method we find in caricature. There is, obviously, some difference: caricature does not apply concretization; it does not cover the distance between the prolonged nose of a man, through a middle sized one, to the portrait with an ordinary nose in the centre of the man's face. Nevertheless, it is a deviation from the same standard: in order to say the truth about a fact it should not be presented as common-sense used to say, but distorted. And doing this, both science and caricature achieve, or at least are able to achieve, the truth. The problem arises as to what kind of truth is arrived at by them.

(b) Classic Notion of Truth and Idealization

It could seem that scientific statements (i.e., the theorems of empirical sciences) are qualified from the viewpoint of the classic definition of truth. However, this conjecture leads to an absurd conclusion that all idealizational statements are true in the classic sense, since they are emptily satisfied by actual objects. Let us say, not only the following law is true in the classic sense:

(13) if $ff(x)$ and $R(x) = 0$, then $F(x) = w(x)$

where $ff(x)$ means that x is a freely falling body; R is the resistance of the air; F is the force applied to a given body; and w is the weight; but also the 'counter-law' of the form:

(14) if $ff(x)$ and $R(x) = 0$, then $F(x) \neq w(x)$.

Note also that from the viewpoint of the classic concept of truth the laws of Galileo, Newton or Einstein are all true, not because they are better than the previous ones, which is what physicists say, but because they are vacuously satisfied, i.e., on purely formal grounds. If one accepts both the classic definition of truth and the thesis that idealization is the basic method of science, it is difficult to understand of what scientific progress consists.

The same holds for caricature: the contents of caricature cannot be qualified from the viewpoint of the classic definition of truth. The statement 'if A has such a long nose, then he looks like this . . . ' presenting the cognitive contents of caricature is emptily satisfied, quite like the statement 'If A has hands as long as that, then he looks like this . . . .' These statements are therefore equivaluable from the point of view of the classic definition of truth (the same holds for (13) and (14)). But if A really has a long nose, it is only the former which will express the contents of a good caricature, in the same way as it takes only the statement (13) to express the law in case of free fall.

(c) *The Thesis of Phenomenalism and the Classic Definition of Truth*

The question arises, why does the classic defintion of truth neglect the basic type of scientific laws? The suspicion arises that this is the case for the same reason which makes modern philosophy of science neglect the method of idealization: both cases result from ontological phenomenalism (see Chapter 7, Section 1(*g*)).

As we have seen, if it is assumed that all the factors influencing a factor in question are equally essential for it, then there is no need to introduce idealizing conditions which disregard some of them. This need becomes powerful whenever the thesis of essentialism is accepted, i.e., if the essential stratification of the factors determining a given one is allowed; then factors believed to be secondary are eliminated with the aid of idealizing conditions. Since the classic definition of truth qualifies all the idealizational statements equally as (vacuously) true, no matter whether they pretend to describe the inner level of the essential structure of the magnitude in question or not, it is not adopted to the idealizational vision of science. For the latter treats the problem of differentiation of idealizational statement as the basic one. Next, since the idealizational vision of science presupposes ontological essentialism, it may be stated that the classic definition of truth presupposes ontological phenomenalism. In fact, if the thesis of ontological phenomenalism is assumed, then only factual statements may be formulated and, at the same time, the classic definition of truth qualifies non-trivially only the factual statements. The classic concept of truth is, then, satisfactory only in the realm of factual statements.

However, science employs the method of idealization as a fundamental one, and the latter presupposes ontological essentialism. Therefore, the classic concept of truth is not assumed in empirical sciences. The problem arises of how one can construct the concept of truth which is based on essentialism and which would be suitable for evaluation of idealizational statements.

(d) The Essentialist Concept of Truth: The Main Idea

The basic idea of the concept of truth which would be adapted to idealizational statements seems to be this: a statement is increasingly true whenever its image of the essential structure comes closer to the actual essential structure of the magnitude in question, and whenever the image of nomological structure assumed by the statement comes closer to the actual nomological structure of the magnitude described in this statement. In short, statement A is more true than statement B, if A presupposes the essence of a given phenomenon more properly.

Let the essential structure of the magnitude F and its nomological structure assume the following form:

$$\begin{aligned} S_F\colon\ &(1) \quad H, \\ &(0) \quad H, p, \\ N_F\colon\ &(1) \quad f_1, \\ &(0) \quad f_1, \{g_0\,, h_0\}. \end{aligned}$$

The following idealizational statement:

(15) if $U(x)$ and $p(x) = 0$, then $F(x) = f_1(H(x))$

is more essentially true, according to the above intuition, than the idealizational statement:

(16) if $U(x)$ and $q(x) = 0$, then $F(x) = f_1(H(x))$.

The former assumes the image of essential structure identical with the structure itself:

$$I_1(S_F)\colon \begin{array}{l}(1)\ H,\\ (0)\ H, p,\end{array}$$

while the second image of the structure is only basically similar to S_F:

$$I_2(S_F)\colon \begin{array}{l}(1)\ H,\\ (0)\ H, q.\end{array}$$

Note that both statements (15) and (16) would be true in the sense of the classic definition of truth. Note also that from the above intutition it follows that the proper concretization of a given idealizational statement is more essentially true than this statement, because the image of the nomological structure which is presupposed by a proper (i.e., revealing the actual directional and correctional functions) concretization is a richer one than that presupposed by the idealizational theorem in question. For instance, (15) presupposes the following image of the nomological structure for the magnitude F:

$$I_1(N_F)\colon (0)\ f_1,$$

while its proper concretization:

(17) if $U(x)$ and $p(x) \neq 0$, then $F(x) = f_0(H(x), p(x))$
$= g_0[f_1(H(x)), h_0(p(x))]$

presupposes a more complete one:

$$I_2(N_F)\colon \begin{array}{l}(1)\ f_1\\ (0)\ f_1, \{g_0, h_0\}.\end{array}$$

(e) The Essentialist Concept of Truth: Some Partial Explications

The explication of the above idea is not of a special importance for this work which deals rather with methodological problems than with strictly epistemological ones. Hence I shall present only some explications of special cases of the essentialist concept of truth employing some intuitions present in the tradition of the Marxist epistemology. My aim here is to show that idealizational theory of science forces some non-classical ideas concerning the nature of truth.

There is a concept of absolute falsity in the traditions of the Marxist epistemology. It could be explicated as follows: a statement t_F (i.e., the statement of the form 'if . . ., then $F =$ . . .') is *absolutely false*, when both the image of the essential structure $I(S_F)$ and the image of the nomological structure $I(N_F)$ assumed by this statement have no elements in common with the structures S_F and N_F, respectively. Let the essential and nomological structure assume the same form as above:

$$\begin{array}{ll} H & f_1 \\ H, p & f_1, \{g_0, h_0\}. \end{array}$$

Then the following statement is, for instance, absolutely false:

(18) if $U(x)$ and $q(x) = 0$, then $F(x) = k(r(x))$.

Another concept is that of partial truth: t_F is *partially true* if in the image $I(S_F)$ there is at least one factor considered to be secondary which is secondary in fact, i.e., occurs as a secondary one in S_F, whereas the image $I(S_F)$ is basically different from S_F, and in the image $I(N_F)$ there is a dependence (correctional or directional function) being assigned to this factor which is an actual dependence, i.e., occurs in the nomological structure N_F, whereas $I(N_F)$ assumed by t_F is basically different from N_F (i.e., it differs with respect to the regularity in question). For instance the statement:

(19) if $U(x)$ and $p(x) \neq 0$, then $F(x) = g_0[k(r(x)), h_0(p(x))]$

is a partial truth if the form of S_F and N_F from the last example is accepted.

The statement t_F is said to be *relatively true* if the images $I(S_F)$ and $I(N_F)$ assumed by the statement are basically similar to the structures S_F and N_F correspondingly, but not identical with them. For instance, the statement

(20) if $U(x)$ and $q(x) = 0$, then $F(x) = f_1(H(x))$

is a relative truth, if the same form of S_F and N_F is presupposed. And the absolute truth is a statement t_F which presupposes the images $I(S_F)$ and $I(N_F)$ identical, respectively, with the structures S_F and N_F. For instance, under accepted schemes of S_F and N_F the following statement will be absolutely true:

(21) if $U(x)$ and $p(x) \neq 0$, then $F(x) = g_0[f_1(H(x)), h_0(p(x))]$.

It follows from this explication that a statement absolutely true is a factual statement, while relatively true statements or partially true ones can be idealizational or factual in nature.

The above–mentioned partial explications should not be viewed as the essentialist theory of truth. On their basis one may only establish conditions of adequacy for such theory, according to which absolute falsities are to be less essentially true estatements than partial truths, the latter are to be less essentially true statements than relative truths, which, in turn, are to be less essentially true than absolute truths. Those partial explications are based on numerous simplifying assumptions which should be removed: that factors influence each other within their entire range (due to this condition the role of the realistic assumption $U(x)$ is not accounted for in the above explications); that there occur on all levels of essential structures (and their images) only singular factors, not sets of equi–significant factors; that an image of the essential structure contains the same amount of levels as the essential structure it reflects; etc. Some of the assumptions of the kind have been removed in the paper by Nowakowa (2) where some more general definition of the relation *being more essentially true than* has been elaborated.

(f) Science and Art

The main ideas suggested above the following: idealizational statements cannot be evaluated from the point of view of the classical definition of truth which deals with factual statements only; it is essential concept of truth which would be able to evaluate idealizational statements. These ideas can be perhaps considered to hold not only for science: caricature is an example which points out that art, too, employs the method of idealization (or a counterpart of it) and hence presupposes the essentialist concept of truth. When a writer collects the elements of life-stories of many persons in order to choose some elements of them and construct in this way the life-story of his protagonist, then it is evident that the evaluation of the novel in question from the point of view of the classic definition of truth is meaningless (as was the case with idealizational statements). The protagonist of a novel is a construct resembling a material point; he has the properties that the author of the novel thinks significant for the category of people described, and has no properties which the author thinks to be minor to them. This example suggests that science and literature, and perhaps art in general, have a common nature: they aim at truth in the essentialist sense of the term. The difference that holds between them is the difference of means applied towards realization of the same effect.

CHAPTER 9

DEDUCTION AND MODELLING (MODEL II)

1. THEORY AS A SEQUENCE OF MODELS

(a) Removal of p_5

Model I consists of methodological hypotheses which state how the ideal researcher, characterized by the assumptions p_1–p_5, carries out his basic research objectives: how he formulates laws, constructs theories and tests them, and explains facts. Conditions p_1–p_5 are undoubtedly idealizing in nature, hence to explain the structure of actual scientific activities more closely it is necessary to remove step by step these conditions and to correct the image of scientific investigations presented in model I.

Let us remove first condition p_5 and assume realistically that our researcher always assumes some idealizational theories in the sense of model I (i.e., simple idealizational theories) of a set of factors, i.e., his theoretical background is non-empty. In model I our investigator's *theoretical* background was assumed, due to condition p_5, to be empty: his initial knowledge, i.e., the knowledge presupposed in theory-construction, contained only empirical registrations and an ontological perspective. In model II we adopt realistically that the ideal investigator possesses a theoretical knowledge consisting of simple idealizational theories of the factors $F_1, \ldots, F_w$ and with this theoretical background in mind he starts to build explanations of facts of the type F.

In less abstract conditions of model II it turns out that the investigator under consideration need not formulate a separate simple idealizational theory of the factor F but he may try to deduce idealizational statements concerning the factor in question from the body of theoretical knowledge. In other words, he need not apply the strategy 'try and test' but he can apply another one: 'try to deduce the theory in question from the theories which are already verified; if this turns out to be impossible, then come back to the first strategy'.

(b) Complex Idealizational Theory

The investigator in question starts from finding out whether the idealizational law of the factor F can be deduced from those statements which have been formulated and verified earlier (and which he accordingly holds to be scientific laws). He thus constructs a *complex idealizational theory* of the factor F. That theory is a sequence of *theoretical models.* that is, pairs (Z^m, T^m), ..., (Z^1, T^1), (Z^0, T^0), which satisfy the following conditions: (1) Z^i is the set of idealizational statements (each based on the same i idealizing assumption) which are logically independent of each other; T^i is an idealizational statement, based on the same i assumptions, which follows from Z^i and describes the dependence of F; (2) Z^i is the set of idealizational laws; (3) Z^i is the union of a non-empty set R^i and a set N^i, (4) for every statement t in R^i there is statement t' in R^{i+1} such that t is its concretization; (5) N^i contains those statements which describe the dependence of those adventitious factors which are taken into account after the removal of the $(i+1)$th idealizing condition (for $i = 0, \ldots, m$). Z^i thus stands for the *assumptions* of the ith theoretical model, T^i is the *solution* of that model, and elements of N^i are *additional assumptions* of the model.

As an example of a theory of this type let us consider the multiplier theory of economic growth (Klein (1), pp. 360–368). It will be shown here already in its reconstructed form. The theory consists of two models, the simple and the expanded one. Symbolism: Ec – the set of economic systems, F – imports, E exports, Y – national product, C – consumption, I – investments. c_0, i, n, m – constants. Consider a very simple form of that theory when it is based on a single idealizing assumption only:

(1) if $Ec(x)$ and $F(x) - E(x) = 0$, then $C(x) = c_0 + mY(x)$.
(2) if $Ec(x)$ and $F(x) - E(x) = 0$, then $Y(x) = C(x) + I(x)$.

Now (1) and (2) yield

(3) if $Ec(x)$ and $F(x) - E(x) = 0$, then $Y(x) = (1/(1 - m))I(x) + c_0/(1 - m)$.

On removing the idealizing assumption we obtain an expanded model for an open economic system:

(4) if $Ec(x)$ and $F(x) - E(x) \neq 0$, then $C(x) = c_0 + mY(x)$,
(5) if $Ec(x)$ and $F(x) - E(x) \neq 0$, then $Y(x) = C(x) + I(x) + E(x) - F(x)$,

(6) if $Ec(x)$ and $F(x) - E(x) \neq 0$, then $F(x) = i_0 + nY(x)$.

These statements yield

(7) if $Ec(x)$ and $F(x) - E(x) \neq 0$, then $Y(x) = \frac{1}{1 - m + n} I(x)$

$$+ \frac{C_0}{1 - m + n} + \frac{E(x)}{1 - m + n} - \frac{i_0}{1 - m + n}.$$

It can easily be seen that (4) and (5) are, respectively, concretizations of (1) and (2), with (4) being a *degenerate concretization* of (1), since the removal of an idealizing assumption is not accompanied by a correction of the subsequent (the correction is empty). And (6) is a new theorem which takes into account the adventitious factor introduced at a lower level of abstraction.

Thus we see that the multiplier theory of economic growth (in its simplified form analysed here) is a complex idealizational theory of the first kind constructed of two models: $(Z^1, (3))$, $(Z^0, (7))$, where $Z^1 = R^1 = \{(1), (2)\}$, $Z^0 = R^0 + N^0 = \{(4), (5), (6)\}$. Thus, in model II a set of universal statements is a scientific theory if and only if it is a complex idealization theory.

2. THE INTERNAL STRUCTURE OF AN IDEALIZATIONAL THEORY

(a) The Internal Structure of Idealizational Laws

The problem arises as to what are the components of an idealizational theory. This problem of the internal structure of an idealizational theory will be first posed with respect to a particular idealizational statement, for instance a law of the form:

(8) if $U(x)$ and $q_1(x) = 0$ and... and $q_m(x) = 0$,
then $F(x) = k_m(B_1(x), ..., B_r(x))$.

This problem consists in the question: What are the methods of forming the statement (8) from the presupposed constructs of a non-sentential nature? The same problem arises, evidently, with respect to the concretization of the statement (8):

(9) if $U(x)$ and $q_1(x) = 0$ and ... and $q_{m-1}(x) = 0$ and $q_m(x) \neq 0$,
then $F(x) = k_{m-1}(B_1(x), ..., B_r(x), q_m(x))$

and, hence, with respect to the simple idealizational theory. It may be then posed also in connection with a complex idealizational theory which is our aim here, because of the intentional applications of the results we shall arrive at. This is why the problem of the internal structure of idealizational theorems is considered within model II, not within model I. Note also that in some points the location of this problem is not yet suitable – the notion of approximation (see Chapter 2, Section 3) will be applied here which our ideal researcher is not yet allowed.

Let us start with a discussion of the semantic character of idealizational statements. It will be convenient to join this discussion with criticism of the positivist grasp of laws of science in Nagel's classic *The Structure of Science*. According to Nagel,

> for an unrestricted universal to be called a law it is a plausible requirement that the evidence for it is not known to coincide with its scope of predication (Nagel (1), p. 63).

The "evidence for a law" is interpreted as those facts which satisfy that law non-vacuously, i.e., satisfy both the antecedent and the consequent of that law:

> In calling a statement a law, we are apparently asserting at least tacitly that as far as we know the examined instances of the statement do not form the exhaustive class of its instances (*ibid.*, p. 63).

Yet in the case of idealizational laws this condition is precisely not satisfied. The scope of application of an idealizational law is exactly the same as the "evidence for that law" (in Nagel's sense): both sets are empty. This is so because in the light of accepted knowledge there is no such fact which would satisfy both the antecedent (i.e., the idealizing assumptions) and the consequent of an idealizational law of type (8). Hence the scope of the application of an idealizational law is empty. But then it follows therefrom that there is no fact, registered by the researcher in question, which would satisfy both the antecedent and the consequent of (8); hence the "evidence for that law" is also an empty set.

Although the above condition which is imposed upon the concept of law of science eliminates idealizational statements from the set of such laws, Nagel many a time quotes such statements as examples of laws in his sense of the term. But Nagel is inconsistent not only on this point. In these cases he is also inconsistent on another point, which – by the way – is much more important. On many occasions he quotes idealizational laws as examples of experimental laws (in his terminology). These are laws which formulate relations between things or features of things that are commonly said to be themselves observable, whether with unaided senses or with the help of special instruments of observation (*ibid.*, p. 79).

And then he continues:

In consonance with this terminological stipulation and the distinction covered by it, the law that the pressure of an ideal gas whose temperature is constant varies inversely with its volume (*ibid.*, p. 80).

would be classed as an experimental law. This conclusion, however, is absolutely unacceptable. Boyle's law, to which Nagel here refers, is an idealizational law, since it is to hold for ideal gases, i.e., as Nagel says on another occasion (*ibid.* p. 360) gases in which the volume of particles (v_w) equals zero and the forces of interaction between particles (p_w) are also all equal to zero. Thus, the full formulation of this law is as follows:

(10) if x is a portion of a gas and $v_w(x) = 0$ and $p_w(x) = 0$, then, if $T(x) = \text{const}$, then $p(x)v(x) = \text{const}$.

where T, p, v stand, respectively, for the temperature, the pressure, and the volume of a portion of a gas. For this definition of the concept of an ideal gas, Boyle's law can also be recorded thus:

(11) if x is a portion of an ideal gas, then, if $T(x) = \text{const}$, then $p(x)v(x) = \text{const}$.

This formulation shows evidently that Boyle's law cannot be an experimental law, since

the rubric 'experimental law' signifies simply that a statement so characterized formulates a relation between things (or traits of things) that are observable. . . . and that the law can be validated . . . by controlled observation of the things mentioned in the law (*ibid.*, p. 80).

But how can we observe ideal gases if it is impossible for any gas to be ideal since there is no gas whose particles are material points that do not interact with one another?

Yet, in the book under consideration, idealizational laws are often quoted as examples of experimental laws. Thus, for instance, Nagel refers to Boyle–Charles's experimental law for ideal gases, the experimental law for a mathematical pendulum (*ibid.*, p. 79), the experimental law for the radiation of an ideally black body (*ibid.*, p. 86), etc. Nevertheless, it seems self-evident that all these idealizational statements are not experimental laws in Nagel's sense as they contain idealizational concepts which are not applicable to observable objects, since they are not applicable to any real objects at all. The question arises, accordingly, whether idealizational laws may be included in the second category of laws as singled out by Nagel, namely theoretical laws (theories).

Now

> . . . the fundamental assumptions of a theory constitute a set of abstract or uninterpreted postulates, whose constituent non-logical terms have no meaning other than those securing to them by virtue of their place in the postulates, so that the basic terms of the theory are 'implicitly defined' by the postulates of the theory (*ibid.*, p. 91).

To put it briefly, theoretical laws are semantic postulates for the theoretical terms they contain. The guess that Nagel would class certain statements on ideal types as theoretical laws is confirmed by his analysis of Newton's laws of motion:

> . . . the axioms of motion must be supposed to be formulated for so-called 'point-masses' – for bodies whose masses are in theory concentrated at a 'point'. The application of the axioms to the motions of actual physical bodies, which are clearly not point-masses, thus assumes an extension of the fundamental theory to cover the motions of systems of point-masses that are subject to more or less rigid mutual constraints. (...) But the facts just noted make it evident that the axioms of motion are *theoretical* statements in the sense of theory previously discussed; they are not statements about relations between experimentally specified properties, but are postulates implicitly defining a number of fundamental notions that are otherwise left unspecified by the postulates of the theory (*ibid.*, p. 160).

Hence, should idealizational statements of type (8) be theoretical laws in Nagel's sense, then they should be semantic postulates for the concepts they contain, in particular the term F. Consider now the example given by Nagel in order to find out whether this is fact in so. If f stands for the external forces working on a body, m, for that body's mass, and v, for that body's velocity, then Newton's first axiom is as follows:

(12) if x is a material point and $f(x) = 0$, then $m(x)(dv(x)/dt(x)) = 0$.

Consider now whether (12) in fact is a semantic postulate for the term 'point-mass'. Now it seems that this is not so. First, the definition of point-mass current in physics is: "Point-mass in mechanics is a body whose dimensions and shape may be disregarded in a given problem" (Frish–Timoreva (1)). In other words, x is a point-mass if and only if x is a body and the dimensions of x all equal zero. This is an equational definition which imparts complete semantic interpretation to the term 'point-mass'. Hence the joining of any other semantic postulates, e.g., in the form of 'theoretical laws', for that term is superfluous as long as this definition holds in physics. Secondly, Nagel's claim that statements of type (12) are semantic postulates results in the conclusion that the meaning of such terms as 'point-mass' is modified each time the old theory is rejected or a new theory is accepted. This, however, is clearly at variance with the uniform interpretation of the term 'point-mass' in the language of physics despite all changes in the theories accepted.

It follows from the above that idealizational laws such as (11) and (12) are not analytic statements, and hence are not theoretical laws in Nagel's sense. This is also indicated by the fact that in empirical sciences idealizational laws are verifiable. Thus, for instance, (11) can be verified by a strict concretization following which less abstract laws, valid for fewer idealizational assumptions and containing corrections which account for the deviation of real gases from ideal gases, are derived from (11). Likewise, (12) can probably be verified by an approximate concretization; we formulate its approximate analogue which states that real bodies which come 'sufficiently close' to point-masses and which are subject to the working of forces which are 'sufficiently balanced', satisfy the approximate equation $m(x)(dv(x)/dt(x) \approx 0$. This is so because it is assumed in physics that the replacement of a body by a point-mass is a justified approximation not only if the dimension of that body are small as compared to other distances taken into account in a given problem, but also we are interested only in the motion of the centre of gravity of that body. On the other hand, verification of idealizational laws would be meaningless if the latter were in fact theoretical laws in Nagel's sense, i.e., if they were semantic postulates for the concepts they contain.

The conclusion from the above is that not only do idealizational laws not meet the general characteristics of laws of science, as given by Nagel, but – contrary to his examples – they are neither experimental nor theoretical laws. Hence they have no place in Nagel's typology of scientific statement.

(b) The Internal Structure of Scientific Theories

All this shows that the image of the structure of a theory as consisting of the following three components is inadequate:

(1) an abstract calculus that is the logical skeleton of the explanatory system, and that implicitly defines the basic notions of the system; (2) a set of rules that in effect assigns

an empirical content to the abstract calculus by relating it to the concrete materials of observation and experiment; and (3) an interpretation or model for the abstract calculus, which supplies some flesh for the skeletal structure in terms of more or less familiar conceptual visualizable materials (Nagel (1), p. 90).

If this image of a theory as consisting of (1) an abstract calculus, (2) correspondence rules (co-ordinating definitions), and (3) rules of semantic interpretation, is to be an adequate reconstruction of the structure of theories to be found in empirical sciences, then it must also cover the structure of idealizational theories, and that structure is more intricate.

The structure of idealizational theories is, roughly speaking, as follows. In empirical sciences, mathematical formulas are always treated as already given a semantic interpretation in the abstract domains of mathematical entities as singled out by mathematicians themselves. A physicist who takes into consideration a mathematical formula, e.g.,

(13) $y(dn/dz) = 0,$

does not think about its semantic interpretation; he accepts the interpretation used by the mathematicians, in accordance with which the above formula contains numerical variables and also symbols of multiplication, equality, derivative, zero. His task is merely to give that formula a *substantial interpretation*, i.e., to assign to the variables sets of values of appropriate physical magnitudes, and to the mathematical operations, certain physical relations between such magnitudes. The adoption of *rules of substantial interpretation* which would assign to mathematical entities real physical objects and their properties would in most cases yield false statements about the world around us. Should we, for instance, adopt the rules by virtue of which y would range over the set of numerical measures of physical objects, n would range over the of numerical measures of velocities of such objects, and z would range over the set of numerical measures of periods of physical time, and should we relabel the variables, y, n, z thus interpreted as m, v, t, which are symbols in current use in physics, we would obtain the factual statement

(14) if x is a physical object, then $m(x)(dv(x)/dt(x)) = 0,$

which however, is a false factual statement. But if we adopt the idealizing assumptions which simplify facts (suppose that only two such assumptions come in question):

(p_1) x is a point-mass,

(p_2) $F(x) = 0.$

then we can – or at least this is what the physicists have done – risk the supposition that formula (13) is true for the factual interpretation that assigns appropriate variables to the physical magnitudes mentioned above, magnitudes which characterize the ideal types defined by the assumptions p_1 and p_2. In other words, we may risk the formulation of a physical hypothesis (12) which has been just constructed by the assignment of variables to physical magnitudes, combined with the adoption of idealizing assumptions p_1 and p_2. This hypothesis is next subject to concretization, perhaps an approximation (see Chapter 2, Section 3) following which we obtain an approximate statement that refers to real objects:

(15) if x is a physical object that is "sufficiently small" and $F(x) \approx 0$, then $m(x)(dv(x)/dy(x)) \approx 0$.

It is exclusively extra-observation terms that occur in (15). Hence we have to derive from (15) certain observation statements. Thus the structure of a simple idealizational theory, theory that consists of a single statement only, is as follows: The theory includes the mathematical formula (13); appropriate rules of substantial interpretation and idealizing assumptions which determine the idealizational law (12) and the approximation law (15).

This analysis can also well be applied to a theory consisting of many statements, such as Newtonian mechanics. Instead of (13) we could take into consideration the three formulae assumed by the Newtonian axioms of motion, appropriate rules of substantial interpretation, idealizing assumptions, etc. Thus the internal structure of an idealizational theory consists of: a set of formal formulas; a set of rules of substantial interpretation; a set of initial idealizational laws preceded by an appropriate set of idealizing assumptions; sets of those idealizational statements which are successive concretizations of the initial laws; a set of those factual statements which are approximate concretizations of previous idealizational laws. This, as can be seen, differs from the structure of scientific theories as interpreted by Nagel.

The question can be posed as to whether the formal formulae adopted in the construction of a scientific theory in the field of empirical sciences are semantic postulates of the language of mathematics (or their consequences), in other words, whether they are mathematical theorems. The answer differs from cases to case. In ordinary empirical theories, such as

Newtonian mechanics, this is not so: formula (13) is not a mathematical theorem, but is just a statement formulated in the language of mathematics. But sometimes the set of those formulae assumed by an empirical theory consists of mathematical theorems. In those cases we might, perhaps, speak about applied mathematics (or its part), as distinct from those cases in which the mathematical background of a theory consists of statements formulated in the language of mathematics, but not of mathematical theorems. This problem will now be analysed for the relationship between pure and 'physical' geometry.

(c) Pure and Applied Geometry

Nagel singles out

two general ways of employing geometry within experimental science. (i) The first way (...) consists in specifying independently of Euclidean geometry certain edges, surfaces, and other configurations of material bodies, and then showing that as a matter of observed facts the things so specified conform to the Euclidean axioms within the limits of experimental error. (ii) The second approach consists in using the Euclidean postulates as implicit definitions so that no physical configurations (whether discovered or deliberately contrived) are called 'points', 'lines', and so on, unless they satisfy the postulates within certain limits of approximation (Nagel (1), pp. 225–226.)

On the first alternative – Nagel says – Euclidean geometry is a system of contingent, *a posteriori* statements concerning spatial properties of bodies antecedently classified and named. On the second alternative, Euclidean geometry is a system of a priori rules for classifying and naming such properties (*ibid.*, p. 231).

And later he goes on to say that

On the first alternative the rules for constructing the figures designated as planes, straight lines, etc., are *a priori*, and the Euclidean statements are empirical. On the second alternative the Euclidean postulates are *a priori*, and the assertions that certain figures (constructed or identified in accordance with special rules), are planes, straight lines, etc., are empirical (*ibid.*, p. 233).

Now it seems that neither of these methods conforms with the procedure used in practice by physicists. To prove this we shall consider Euclid's fifth axiom:

(16) Only one straight line can be drawn parallel to a given straight line through a point not on this line.

If the first method is used, then the following definition (whose actual correctness is inessential here) is adopted:

(17) (a) the photon is taken to be a physical point,

(b) the trajectory of a photon is taken to be a physical straight line.

It is then stated that the empirical correctness of the statement that only one physical straight line so conceived and parallel to another straight line so conceived can be drawn through a physical point so conceived and lying not on this line, testifies to an 'approximate truth' of axiom (16). But the saying that (16) is 'approximately true' can only be metaphorical in nature as long as we do not indicate a system of logic which would validate such a statement. This is not done by Nagel who, according to what can be inferred from his text, assumes ordinary two-valued logic. But if so, then we have to conclude that the geometrical statement (16) is true by definition for the ordinary, mathematical, interpretation of the terms *point* and *straight line*. It may not be said that it is satisfied approximately, because – in the light of the logic assumed by Nagel – that would be meaningless. We may only formulate the statement:

(18) only one physical straight line can be drawn parallel to a given physical straight line through a physical point not on this physical straight line,

which is perhaps true, but does not provide any grounds for deciding about the logical value of (16). This is so because (18) is merely constructed after (16). That no logical relation holds between the two can easily be realized when we note that (18) differs from (16) by having the terms 'physical point' ('physical straight line') in those places where 'point' ('straight line') occur in (16). Likewise, the statement 'every swan is white' can be formed from the statement 'every raven is black' by replacing in the latter the appropriate extra-logical expressions, which, however, does not provide any grounds for concluding about the logical value of the latter statement from that of the former one, as there is between them no relation whatever that would validate such inference.

The situation is similar in the case of the second method. Here (16) is treated as a geometrical theorem, and hence as a semantic postulate for the terms 'point' and 'straight line', while the following empirical hypotheses are made:

(19) (a) the photon is a point,
(b) the trajectory of a photon is a straight line.

These hypotheses are supposed to make it possible to state (16) about the physical world. But they are false for the same reasons for which a physical point (physical straight line) as understood in definition (17) is not a point (straight line). And to say that these hypotheses are 'satisfied approximately' is untenable in the light of ordinary logic.

It seems, however, that we can try to reconstruct the method of applying geometry in empirical sciences without resorting to the formulation 'satisfied approximately'. To do so we have to assume that what is usually termed 'physical geometry' is an idealizational theory in the sense of what has been said above. It is also such a theory that its 'formal background' includes geometrical theorems, i.e., semantic postulates of the language of geometry, or their consequences.

We shall consider the example already analysed above in order to bring out the differences between this approach and the methods suggested by Nagel. The physicist starts from statement (16) by treating it as a geometrical theorem; he accordingly interprets the terms 'point' and 'straight line' in the way provided by that theorem. He also lays down rules of substantial interpretation such as these given below:

(20) (a) a point has its analogue in a point-mass.
(b) a straight line has its analogue in an edge of an ideally rigid body.

Obviously, reference to the concept of point-mass is possible only after the adoption of the idealizing assumption that postulates the corresponding ideal type; the same holds, by analogy, for the concept of an ideally rigid body. Making these assumptions and adopting rule (20) the physicist formulates the following idealizational law that is in the field of classical mechanics.

(21) if x is a point-mass and if y is an edge of an ideally rigid body, and if x does not lie on y, there is exactly one edge z of an ideally rigid body such that x lies on z and z is parallel to y.

The difference between the passage from (16) to (21), and the passage from (16) to (18), as postulated by Nagel's first method, is that in Nagel's case an 'approximate truth' of (16) is inferred from the truth of (18), whereas our physicist accepts (16) regardless of whether or not he hypothetically accepts the idealizational statement (21); if he does, then this fact is for him no argument for the acceptance of (16), which he accepts anyway as an analytic statement.

Next the physicist under consideration accepts principles of approximation which state that, for his purposes.

(22) (a) a photon is a 'sufficient approximation' of a point-mass,
(b) the trajectory of a photon is a 'sufficient approximation' of an edge of an ideally rigid body.

These principles enable him to assert (18) by virtue of approximation: (18) is now introduced as an approximation of the idealizational law (21), and hence refutation of observational consequences of (18) would require rejection of (21), unless we reject some of the statements (22). While (21) is in the sphere of idealizational 'physical geometry', (18) is in the sphere of 'factual physical geometry'. The difference between Nagel's first method is that the truth of (18) is not a basis for the assessment of the logical value of the mathematical statement (16). The difference between Nagel's second method is that it is not claimed about the statements (22) that they are 'approximately true', and thus the principles of two-valued logic are not impaired. It is only the objects referred to in those statements that are their own 'approximations'; but this formulation can be given a clear interpretation by indicating the admissible deviations (in size) of one from the other. The principles of ordinary logic are not infringed, since the statements (22), as soon as the limits of admissible approximation are defined, prove to be either true or false (see Chapter 10).

As can be seen from the above, 'applied geometry' covers pure geometry, the appropriate rules of substantial interpretation, idealizational 'physical geometry' and factual 'physical geometry'. Its peculiarity consists only in the fact, perhaps in analogy to other branches of 'applied mathematics', that pure geometry, which forms its 'formal background', consists of semantic postulates of the language of mathematics, and not just of statements formulated in that language.

3. MODELLING

When the enlarged form of an idealizational theory is considered within model II, some actual phenomena of scientific activity may be explained. One of them is the procedure of modelling which leads to building a new theory on the basis of the previous one.

Given two universal statements in the form:

(23) if $z_0(x)$ and ... and $z_n(x) = 0$, then $A(x) = f(B_1(x), \ldots, B_p(x))$,

(24) if $s_0(x)$ and ... and $s_r(x) = 0$, then $C(x) = f(D_1(x), ..., D_p(x))$,

they are said to be *formally analogical* relative to a *formal formula*:

(25) $a = f(b_1, ..., b_p)$

if the numerical variables a, b_i range over the sets of numbers which are, respectively, both the sets of values of A, B_i and the sets of values of C, D_i (for $i = 1,..., p$). If (23) is an idealizational statement, we say that (24), which may then be either idealizational or factual, *formally models* (23). Thus, for instance, Coulomb's law formally models Newton's law of gravitation under the formal formula: $a = b.\ b'/c^2$; the idealizing assumptions of these laws differ, but their consequents fall under the same formula.

A *formal model of the first order* of a theory (Z^k, T^k), ..., (Z^0, T^0) is a theory (S^n, L^n), ..., (S^0, L^0) which satisfies the following conditions: (1) there is a one-to-one correspondence between every theorem t in Z^k and a theorem t' in S^n, t' being formally analogical to t; (2) there is in S^{n-1} a theorem which is not formally analogical to any theorem in Z^{k-1}. Note that it follows from (1) that T^k is formally analogical to L^n. Hence, if at one time the solution T^k of the assumptions Z^k is unknown, then if we know the solution L^n of the assumption S^n, we can find the theorem T^k, which must be formally analogical to L^n. But it follows from (2) that the possibilities of making use of a formal model of the first order end at this point: the analogy does not extend to first-order concretizations, as not all concretizations of theorems which are elements of S^n are formally analogical to concretizations of theorems which are elements of Z^k. Otherwise it is in this case a *formal model of the second order*: (1) holds for Z^{k-1}, S^{n-1} and (2) for Z^{k-2}, S^{n-2}. By analogy, we could introduce the concepts of formal models of higher orders. Obviously, the higher is the order of a formal model of a given theory, the greater is its cognitive significance. For if the formal analogy of the assumptions which occur in the initial models is inherited by remote concretizations, then it is possible to find formally analogical solutions of the successive models.

Let us illustrate the notion of a formal model of a theory with the example of analogy between Marx's theory of reproduction for capitalist economies and one of the theories of general equilibrium of a socialist economy. The former has been already reconstructed (see Chapter 2, Section 2). Let us start then with the reconstruction of the latter.

The starting point is a couple of idealizing conditions:

To simplify our analysis, we assume . . . : the lack of unproductive area, the lack of international exchange, the lack of consumers' savings, the stability of prices (Szefler (1), p. 158).

Under those assumption the following equations are in force:

$$J = A_1 + V_1 + M_1^i,$$
$$K = A_2 + V_2 + M_2^i,$$

where the subscript 1 indicates the sector of the production of the means of production, and 2 denotes the sector of the production of the means of consumption. Moreover, A is the amortization of constant means, V is the fund of consumption, M^i is the part of surplus value designed for new investments, J is the supply of investment goods, K is the supply of consumption goods. From those equations the condition of equilibrium between the production and the consumption is deduced.

Now, we should present the assumptions of the theory in question more exactly:

$$p_0\colon S(x) \text{ (i.e., } x \text{ is a socialist economy)}$$
$$p_1\colon\ M_1^n(x) = 0,\ M_2^n(x) = 0,$$

that is, the fund designed to support unproductive activity in both sectors of an economy x equals zero.

$$p_2\colon E(x) - I(x) = 0,$$

i.e., the difference between the exports and the imports in the economy x equals zero.

$$p_3\colon\ S(x) = 0,$$

i.e., the savings of consumers in the economy x equals zero.

$$p_4\colon p(x) = \text{const.},$$

i.e., prices of products manufactured in the economy x are constant. In this connection, equations cited above can be treated as abbreviations of the following idealizational statements:

(E_1^4) if $S(x)$ and $M_1^n(x) = 0$ and $E(x) - I(x) = 0$ and $S(x) = 0$ and $p(x) = \text{const.}$, then $J(x) = A_1(x) + V_1(x) + M_1^i(x)$

(E_2^4) if $S(x)$ and $M_2^n(x) = 0$ and $E(x) - I(x) = 0$ and $S(x) = 0$ and $p(x) = \text{const.}$, then $K(x) = A_2(x) + V_2(x) + M_2^i(x)$.

Theses (E_1^4) and (E_2^4) are in force only for an idealized economy defined by the conditions p_1–p_4 (and many more conditions of the kind which have been omitted here for the sake of simplicity).

Next, to

make the problem of dynamic equilibrium deeper and to approximate it to reality, we will gradually remove previously accepted simplifying assumptions. First we will introduce unproductive area to our reasoning (Szefler (1), p. 161).

In other words, the author removes the idealizing condition p_1 and corrects the theses, taking into account precisely the unproductive area:

(E_1^3) if $S(x)$ and $M_1^n(x) > 0$ and $E(x) - I(x) = 0$ and $S(x) = 0$ and $p(x) = \text{const.}$, then $J(x) = A_1(x) + V_1(x) + M_1^i(x) + M_1^n(x)$

(E_2^3) if $S(x)$ and $M_2^n(x) > 0$ and $E(x) - I(x) = 0$ and $S(x) = 0$ and $p(x) = \text{const.}$, then $K(x) = A_2(x) + V_2(x) + M_2^i(x) + M_2^n(x)$

One can easily notice that (E_1^3) is a concretization of (E_1^4) and (E_2^3) is a concretization of (E_2^4).

As the following step, the assumption p_2 is removed. It can be achieved in two ways: exports are greater than imports, or the contrary.

From the standpoint of the general equilibrium, if the exports are greater than the imports, then this difference is treated as additional investments abroad (Szefler (1), p. 163).

If the imports are in excess to the exports, then it can be treated as additional investments in a given country (*ibid.*, p. 164).

One can observe that those formulations may be reconstructed in the following manner:

(E_{1a}^2) if $S(x)$ and $M_1^n(x) > 0$ and $(E(x) - I(x)) > 0$ and $S(x) = 0$ and $p(x) = \text{const}$, then $J(x) + (E(x) - I(x)) = A_1(x) + V_1(x) + M_1^i(x) + M_1^n(x)$

(E_{2a}^2) if $S(x)$ and $M_2^n(x) > 0$ and $(E(x) - I(x)) > 0$ and $S(x) = 0$ and $p(x) = \text{const}$, then $K(x) = A_2(x) + V_2(x) + M_2^i(x) + M_2^n(x)$

(E_{1b}^2) if $S(x)$ and $M_1^n(x) > 0$ and $(E(x) - I(x)) < 0$ and $S(x) = 0$ and $p(x) = \text{const}$, then $J(x) = A_1(x) + V_1(x) + M_1^i(x) + M_1^n(x) + (I(x) - E(x))$

(E_{2b}^2) if $S(x)$ and $M_2^n(x) > 0$ and $(E(x) - I(x)) < 0$ and $S(x) = 0$ and $p(x) = \text{const}$, then $K(x) = A_2(x) + V_2(x) + M_2^i(x) + M_2^n(x)$.

From the pairs of statements (E_1^4)–(E_2^4), (E_1^3)–(E_2^3), (E_{1a}^2)–(E_{2a}^2) and (E_{1b}^2)–(E_{2b}^2) the appropriate conditions of equilibrium for a socialist economy are deduced.

Now one may observe that the theory of reproduction from *Capital* (see Chapter 2, Section 2) is a complex idealizational theory in the sense of model II. Comparison of this theory with the theory of general equilibrium of socialist economies demonstrates that the latter is a formal model of the first order of Marx's theory. Compare the schemes of simple reproduction (see Chapter 2, Section 2):

(R_1^2) if $E(x)$ and $Val(x) = \text{const}$ and $M_a(x) = 0$, then $P_1(x) = C_1(x) + V_1(x) + M_1(x)$,

(R_2^2) if $E(x)$ and $Val(x) = \text{const}$ and $M_a(x) = 0$, then $P_2(x) = C_2(x) + V_2(x) + M_2(x)$,

with the statements (E_1^4), (E_2^4). It is evident that under assignment:

$$J \text{———} P_1$$
$$K \text{———} P_2$$
$$A_1 \text{———} C_1$$
$$V_1 \text{———} V_1$$
$$M_1^i \text{———} M_1, \text{ etc.}$$

(E_1^4) is formally analogical to (R_1^2) and (E_2^4) is formally analogical with (E_2^2): realistic and idealizing conditions of these statements are, as pairs, different but their consequents are, as pairs, substantial interpretations of the same formal schemes:

$$a_1 = b_1 + c_1 + d_1,$$
$$a_2 = b_2 + c_2 + d_2.$$

At the same time it is sufficient to compare the schemes of enlarged reproduction (R_1^1) and (R_2^1) (see Chapter 2, Section 2) with the statements (E_1^3), (E_2^3) to observe that there is also formal analogy between them. Therefore, one may conclude that the theories in question are formally analogical at least in the second degree, i.e., that the theory of equilibrium of a socialist economy is a formal model of the second order at least of the theory of reproduction of a capitalist society.

4. CONCLUSIONS

As we have seen, only one aim of our investigator is achieved differently within model II than it had been achieved within model I: building a theory. A theory within model II is defined not only by the relation of concretization but by both the relations of concretization and of entailment:

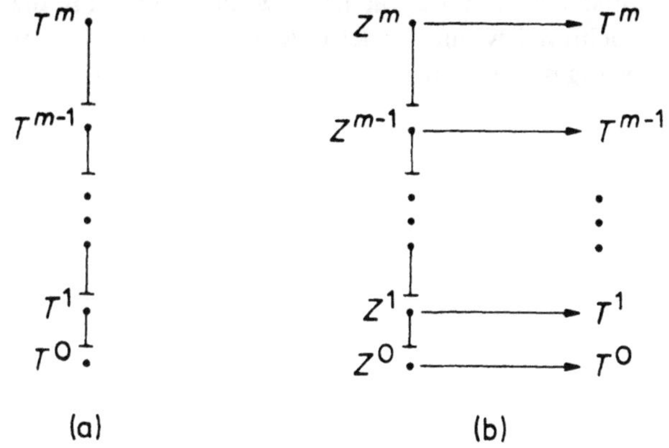

Fig. 4. (a) The structure of an idealizational theory in the sense of model I (i.e., of a simple idealizational theory) is defined by the relation of concretization ⊣ only. (b) The structure of an idealizational theory in the sense of model II (i.e., of a complex idealizational theory) takes into account the additional dimension of actual scientific theories, viz. that of entailment (→). Due to this it reveals not only the role of the procedure of idealization and concretization (as the simple theory does) but also the role of deduction in theory-building.

A complex idealizational theory is a more realistic form of theory than a simple idealizational theory, since the latter accounts for not only the relation of concretization but also the relation of entailment, explaining in this way the role played in theory-construction by the procedure of deduction. The way of explanation within model II has been only slightly modified: in model II, where complex idealizational theories are constructed, links of the chain of explanation must not coincide with elements of a theory (as was the case within model I) but can be taken from the various theoretical models included in a complex theory.

The rest of our methodological hypotheses from model I remains unchanged: the criteria of distinguishing laws within universal statements from idealizational theories are the same both in model I and in model II. And the same holds for criteria of acceptance of theorems: verification through perfect experiment is maintained also in model II. In other words, accounting for the fact that theory-construction always presupposes a theoretical background (not only philosophical and/or empirical) results in the structure of idealizational theories built within model II which, in turn, changes slightly the model of explanation applied within this model. Another scientist's objectives are achieved within model II in the same way as they were achieved within model I. After removing the next assumptions, other changes in the structure of scientific activities will occur.

CHAPTER 10

APPROXIMATION

1. APPROXIMATION THEORY

(a) Removal of Assumptions p_4 and p_3

We have so far assumed the condition p_4 which seems to be rather strong – it is rather strange to expect that a researcher always knows what the effects are of all factors considered by him to be secondary ones for the magnitude in question. The same holds for the condition p_3 of a possibility of perfect experiments – technical capacities of actual researchers are sufficient to *minimize* the influence of *some* factors only.

Therefore the assumption p_4 is removed according to which an ideal investigator knows the influence of adventitious factors, that is, has no doubts as to how to concretize the initial idealizational statements. Since this condition is not always met, we adopt the realistic assumption which states that the researcher in question knows how some of the adventitious factors work, but does not know how the remaining ones do. It follows therefrom that the researcher knows how to concretize his initial statements with respect to a number of idealizing conditions but cannot do that with respect to the remaining ones. We also remove the assumption p_3 and replace it by the realistic condition which states that the researcher in some cases succeeds in minimizing the effect of adventitious factors upon the magnitude he studies, but is never in a position to reduce the effect of all those factors to the zero level.

(b) Approximation Statements

Given an idealizational statement:

(1) if $U(x)$ and $q_1(x) = 0$ and ... and $q_m(x) = 0$,
then $F(x) = f_m(B_1(x), ..., B_r(x))$,

the factual statement:

(2) if $U(x)$ and $q_1(x) \leq a_1$ and ... and $q_m(x) \leq a_m$,

$$\text{then } F(x) \approx_\varepsilon f_m(B_1(x),\ldots, B_r(x))$$

is called an *approximation statement* with respect to (1). Note that (1) a_i is a number in the set of values of the magnitude p_i other than 0, (2) $\varepsilon = k(a_1, \ldots, a_m)$, where $a_i \neq 0$, and k is a function such that the smaller are a_i for the constant remaining numbers of this type, the smaller is $k(a_1, \ldots, a_m)$, (3) $N > \varepsilon$ where N is a constant adopted by the researcher. The function k is the *criterion of approximation*, and the constant N (which the researcher usually wants to have 'sufficiently small') is the *threshold of approximation.* Now (2) states roughly that the closer the real conditions come to the ideal conditions described in the antecedent of the idealizational statement (1), the closer the value of the magnitude under consideration comes to $f(B_1(x), \ldots, B_r(x))$; in any case, the difference between the two cannot exceed the threshold of approximation N, adopted by the researcher in advance.

(c) Complex Approximation Theory

Under presently accepted assumptions a researcher concretizes his initial statements with respect to the working of those adventitious factors which he knows how to take into account, and then applies the approximation procedure: he summarily takes into consideration the working of all the remaining factors; in doing so he merely sees that the discrepancy between his results and the empirical data should not exceed a certain admissible measure. The structure of a *complex approximation theory* is therefore as follows: it is a sequence of theoretical models $(Z^m, T^m), \ldots, (Z^i, T^i)$, (AZ^i, AT^i) where the sequence $(Z^m, T^m), \ldots, (Z^i, T^i)$ has the structure of a complex idealizational theory and the set AZ^i includes the approximations of the idealizational statements which are in the Z^i. Hence in model III now becomes the new form of a theory – that of approximation theory.

2. MEASUREMENT

Within models I and II the structure of measurement has not been considered at all. At this stage of our analysis the conceptual apparatus is rich enough to discuss this.

Let us start with noticing that within the positivist model of science reference is sometimes made to the notion of ideal types, idealizing conditions, etc., in connection with problems of measurement whereas it is clearly not valid in the light of the general positivists assumptions (see Chapter 4). No wonder that this results in inner contradictions in their views on science. Let us discuss this problem considering the example of Nagel's methodological concept.

As we have seen (Chapter 9, Section 2), Nagel's concepts of a law of science and of a theory do not account for the place of idealizations in empirical sciences. What is more, they even exclude the method from a body of scientific procedures. Nevertheless, discussing measurement Nagel says that every method of the measurement of length

> tacitly involves the notion of a rigid body which in theory is isolated from influences that may alter the relative lengths of physical objects, and which by definition therefore has an unchanging length (Nagel (1), p. 255).

But remarks of the kind (*ibid.*, p. 255 ff) result in consequences which are at variance with the general interpretation of the problem of measurement as adopted by Nagel. It turns out, for instance, that the measurement statement on the dimensions W of the concrete object x_0 is not at all in the form of the measurement statement:

(3) $$W(x_0) = kd(s_0)$$

where s_0 is a given measuring rod, d its length and k the number which indicates how many times the measuring rod has been (in fact or in imagination) 'applied' to the object being measured. The object s_0, referred to in the measurement statement which, in agreement with Nagel's formulation quoted above, is in fact formulated in science, is not a physical object, as in (3), but an ideal type of an ideally rigid body. This means, however, that actual measurement statements as formulated by physicists are not in the form of the observational statement (3), but are in the following form:

(4) if s_0 is an ideally rigid measuring rod, then $W(x_0) = kd(s_0)$.

These are hence idealizational, and not observational statements, since ideally rigid bodies, being non-existent, cannot be observed. Thus, Nagel's interpretation results in incoherences in his system.

The question remains how the measurement procedure works as a verifier of hypotheses. Idealizational measurement statements require, as all measurement statement do, a concretization or approximation. To reconstruct this in greater detail we need to reveal the meaning of the concept of a rigid measuring rod. The latter term is being applied to a measuring rod which is not subject to the influence of what are termed differentiating forces, i.e., forces that are not universal in nature, as is, for instance, the force of gravitation, but work differently according to the kind of objects to which they are directed. This being so, an *idealizational measurement statement* might be formulated thus (where F_r stands for the sum of differentiating forces, $A(s_0, x_0)$ means that rod s_0 is applied to object x_0):

(4′) if $F_r(s_0) = 0$ and $A(s_0, x_0)$, then $W(x_0) = kd(s_0)$.

The abolition of the idealizing assumption which occurs in the antecedent of (4′) can follow either as a result of an approximation or as a result of a concretization. In the former case we conclude that the measuring rod we use is 'sufficiently rigid' for our purposes, although, obviously, it is not ideally rigid. Hence the result of the measurement made with a sufficiently rigid rod can be accepted as a 'sufficient approximation' of the real dimensions of the object involved. Thus approximate concretization yields the following factual measurement statement:

(5) if $F_r(s_0) \leq \alpha$ and $A(s_0, x_0)$, then $W(x_0) \approx_\varepsilon kd(s_0)$,

where α stands for the number (which is not always explicitly fixed) that indicates the admissible 'level of deviation' of the real measuring rod from a rigid rod. If it is in fact so that $F_r(s_0) = \beta$, where $\beta \leq \alpha$, then by virtue of this we may accept the statement:

(6) $W(x_0) \approx_\varepsilon kd(s_0)$,

which is an observational measurement statement and can serve to verify hypotheses. If, however, we are concerned with an exact result of measurement, then we have to consider the fact that the rod we use is not rigid, and strictly concretize the idealization measurement statement (4′) by introducing the factual measurement statement:

(7) if $F_r(s_0) = m$ and $A(s_0, x_0)$, then $W(x_0) = kd(s_0) \pm \theta(m)$,

where $m > 0$, and θ is a function which, in the theory assumed, indicates how the differentiating forces distort the measuring rod (for instance, how much they lengthen that rod). If in fact $F_r(s_0) = m$, then we may accept the observational measurement statement

(8) $W(x_0) = kd(s_0) \pm \theta(m)$.

Obviously, results of measurements always are approximate, so that approximation is always involved (at the most it follows concretization), since we as a rule do not take into account rigorously all the differentiating forces.

3. VERIFICATION OF AN APPROXIMATION THEORY

We have already spoken (see Chapter 6) about the principle of idealizational empiricism. At present our construction becomes able to include and explicate ideas underlying the principle in question.

So far, our ideal researcher used to reject his idealizational hypotheses definitively if he obtained, as a result of a perfect experiment, some observations against them. Now after the removal of assumptions p_4 and p_3 the situation has changed, and the principle of idealizational empiricism may be introduced. Let us refer to a pair of Marxian concepts: a fact which *prima facie* negates a statement and a fact which really refutes a statement. A fact which *prima facie refutes* the statement:

(T^m) if $U(x)$ and $q_1(x) = 0$ and ... and $q_m(x) = 0$, then $F(x) = f_m(B_1(x), \ldots, B_r(x))$

is a fact which is at variance with the approximation of T^m:

(AT^m) if $U(x)$ and $q_1(x) \leq a_1$ and ... and $q_m(x) \leq a_m$, then $F(x) \approx_\varepsilon f_m(B_1(x), \ldots, B_r(x))$

but there is such a concretization T^j of T^m:

(T^j) if $U(x)$ and $q_1(x) = 0$ and ... and $q_j(x) = 0$ and $q_{j+1}(x) \neq 0$ and ... and $q_m(x) \neq 0$, then $F(x) = f_j(B_1(x), \ldots, B_r(x), q_m(x), \ldots, q_j(x))$

that this fact is not at variance with the approximation of T^j:

(AT^j) if $U(x)$ and $q_1(x) \leq a_1$ and ... and $q_j(x) \leq a_j$ and $q_{j+1}(x) \neq 0$ and ... and $q_m(x) \neq 0$, then $F(x) \approx_\varepsilon f_j(B_1(x), \ldots, B_r(x), q_m(x), \ldots, q_j(x))$.

In other words, it is a fact which negates the approximation of the idealizational statement in question, but does not negate the approximation of some concretization of this statement. A fact which *really refutes* statement T^m is not only at variance with the approximation of this statement but also with the approximation of its every concretization (or with the final concretization). Thus a fact which *prima facie* refutes an idealizational statement does so only apparently, because the discrepancies between the empirical data and the approximation of that statement can be explained by the working of disregarded adventitious factors. Such a fact shows then that a given idealizational statement may not be approximated yet, because discrepancies between this statement and the empirical data are too great as the disregarded secondary factors work too strongly. Due to this an investigator, according to the principle of idealizational empiricism, does not reject an idealizational statement merely because he has come across a fact which refutes it, but only when it turns out that the fact really negates that statement. Therefore within model III idealizational statements are verified in another manner than in models I and II. First, it is an ordinary experiment minimizing, but not eliminating, the influence of secondary factors which is possible in the more realistic conditions of model III, making experimental verification of idealizational statements possible through approximation only. Second, idealizational statements are verified by observations intended to show whether there are facts which really refute these statements.

4. CONCLUSIONS

The structure of scientific activity noticeably changes within model III. First, theory-construction becomes more realistic than within model II: it accounts for the new dimension of a theory, viz. that of approximation. Actual scientific theories are always approximate descriptions of reality. Model III allows for including this intuition into our construction. Thanks to this, the procedure of approximation is explained. Namely, science primarily resorts to idealization whereas approximation is applied when concretization of idealizational statements is impossible. Approximation is, therefore, a substitutional procedure for concretization. This is clearly seen when the structure of an approximation theory is compared to that of a complex idealizational theory. The former is a more realistic grasp of actual theory-construction in science than the latter – it contains three actual dimensions of real scientific theories: concretization, deduction, and approximation, while the latter contains only two of them.

It follows from what has been said above that the schema of explanation also changes. Now it takes the form:

$$T^m \dashv T^{m-1} \dashv \ldots \dashv T^i \rightsquigarrow\!\!\dashv AT^i \wedge P \rightarrow E$$

where T^m, ..., T^i, AT^i are elements of successive models (being elements

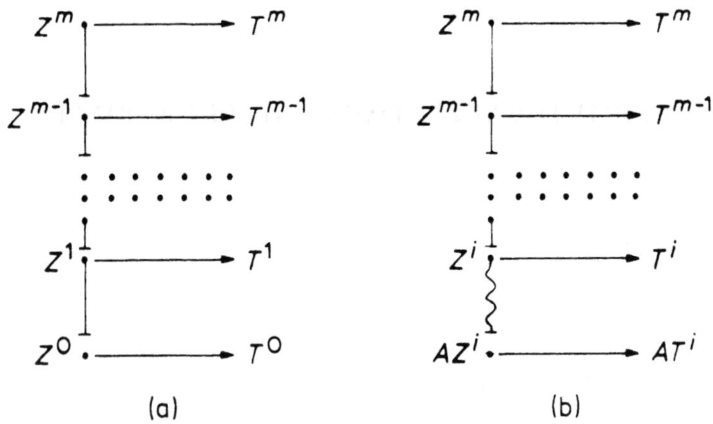

Fig. 5. (a) The structure of a complex idealizational theory (model II) is defined by the relations of concretization and of entailment. (b) The structure of an approximation theory (model III) is defined by the relations of concretization, entailment and of approximation (⇝).

of assumptions of a model or being identical with the solution of it) of the approximation theory; ⇝ stands for the relation of approximation.

The method of verification of a theory also changes. In the place of a perfect experiment an ordinary experiment is applied within model III. It is able to control idealizational statements not directly, as would the perfect experiment, but through approximation only. And the principle of idealizational empiricism begins to be applied – an investigator looks for the facts which really, not *prima facie*, refute hypotheses invented by a member of a scientific group.

CHAPTER 11

SEMI-IDEALIZATION AND PROBABILITY

1. SEMI-IDEALIZATIONAL STATEMENTS

(a) Removal of p_2

We accepted so far assumption p_2 according to which our investigator has no troubles with enumeration of those factors he considers to be essential for the magnitude under consideration. However, this is not the case. There are some factors which are known to be influential for a magnitude under investigation but researchers can neither enumerate them nor mark their influences. After establishing a given image of the essential structure they are supposed to be the *rest* of significant factors for the magnitude in question. Our investigator knows that *there are* some significant factors for a given magnitude besides those shown in his image of the essential structure but he cannot enumerate them. Factors of this kind will be called *interfering factors* for a given magnitude.

(b) An Image of the Essential Structure Open to Interferences

Let us mark the set of interfering factors for F as I^F. According to the assumption p_2, this set is empty for each F and for each investigator. At the present stage of our analysis we assume realistically that for each factor and for each researcher the set of interfering factors, i.e., influence which are known to hold but cannot be identified by an investigator, is non-empty. It follows from this that the concept of an image of the essential structure has to be replaced with a new one denoting a hierarchy of factors that are partly identified and partly not. The hierarchy is of the following form:

$$
\begin{array}{lll}
I^i(S_F): & (m) & B_1, \ldots, B_r \\
& (m-1) & B_1, \ldots, B_r, q_m \\
& \multicolumn{2}{l}{\ldots\ldots\ldots\ldots\ldots\ldots\ldots\ldots} \\
& (1) & B_1, \ldots, B_r, q_m, \ldots, q_2 \\
& (0) & B_1, \ldots, B_r, q_m, \ldots, q_2, q_1 \\
& (0^+) & B_1, \ldots, B_r, q_m, \ldots, q_2, q_1, i^F
\end{array}
$$

and will be called an *image of the essential structure open to interferences.* For instance, let us consider the classic model of demand (Klein (1), p. 23ff):

(1) $$q = \alpha + \beta p + u$$

where q is the demand, p is the price, u is the error, α, β are constant parameters.

Let us try to reconstruct the image of the essential structure underlying the equation. Only one factor is identified as influencing q, viz. the price p. The 'error' follows from this very fact that numerous factors influencing q are not taken into account and their joint influence is estimated as u. Only some examples of these factors are given – rain, temperature, fashion, and so on. All of them – and no one is able to enumerate these factors – are treated as 'accidental interferences' (Klein (1), p. 28). Hence one may state that the considered image looks as follows:

$$I^i(S_q)\colon (0)\ \ p$$
$$(0^+)\ p,\ i^q$$

where i^q is the set of all factors influencing q that are not included into the image of the essential structure in the previous sense:

$$I(S_q)\colon (0)\ \ p.$$

The joint influence of the factors of i^F is estimated as u.

This example enables us to give a more strict definition of the set of interfering factors. Let us note that the set i^F may be identified with the difference of sets P_F and $I(P_F)$, i.e., between the space of essential factors for F and the image of this space in the sense of model I (maintained within models II and III). All the factors influencing F which are not included into the image $I(P_F)$, and therefore into the image $I(S_F)$, are interfering factors for F. This is why the set i^F cannot be enumerated – no one knows exactly what are the factors belonging to the space of essential factors for a given magnitude.

To distinguish the image of essential structure in the sense of model IV (i.e., that open to interferences) from this image in the sense of model III, let us call the latter the *image of the essential structure closed to interferences.*

(c) A Semi-Idealizational Law

The propositional function:

(2) $\bigwedge i^F_x = 0$

will be read-every factor of the set i^F takes on zero value for an object x. This function will be called a *semi-idealizing assumption*: its role consists in elimination of all the interfering factors.

A *semi-idealizational statement* is termed a universal that contains a semi-idealizing condition. If it contains additionally some idealizing conditions the statement is termed a *semi-idealizational statement of the first kind*; if this is not the case the statement is a semi-idealizational statement *of the second kind*.

Within model IV, then, the notion of scientific law is changed, as the term now applies to the statement which is of the form presented within model I (and II and III as well)or of the form:

(ST^m) if $U(x)$ and $\bigwedge i^F_x = 0$ and $q_1(x) = 0$ and ... and $q_m(x) = 0$, then $F(x) = k_m(B_1(x), \ldots, B_r(x))$.

The latter differs from the idealizational law in the standard form T^m only with respect to the semi-idealizing condition (2). Therefore it is called a *semi-idealizational law*. Within model IV scientific laws are now to be idealizational laws or semi-idealizational laws.

Note that the statement which is a semi-idealizational one of the second kind is neither an idealizational one nor a factual one if the latter term is being limited only to those statements that do not possess any kind of counterfactual conditions. Therefore, semi-idealization may be treated as a weakened form of idealization in a broader sense of this term.

2. SEMI-IDEALIZATIONAL THEORIES

(a) Statistical Concretization

Let us assume that our ideal researcher has established the statement ST^m. What he does further is an ordinary gradual concretization of this statement resulting in the sequence of semi-idealizational statements of the first kind:

(ST^{m-1}) if $U(x)$ and $\bigwedge i^F_x = 0$ and $q_1(x) = 0$ and ... and $q_{m-1}(x) = 0$ and $q_m(x) \neq 0$, then $F(x) = k_{m-1}(B_1(x), \ldots, B_r(x), q_m(x))$

....

(ST^1) if $U(x)$ and $\bigwedge i^F_x = 0$ and $q_1(x) = 0$ and $q_2(x) \neq 0$ and ...

and $q_m(x) \neq 0$,
then $F(x) = k_1(B_1(x), \ldots, B_r(x), q_m(x), \ldots, q_2(x))$.

The next statement obtained in this way is a semi-idealizational statement of the second kind:

(ST^0) if $U(x)$ and $\bigwedge i_x^F = 0$ and $q_1(x) \neq 0$ and ... and $q_m(x) \neq 0$, then $F(x) = k_0(B_1(x), \ldots, B_r(x), q_m(x), \ldots, q_2(x), q_1(x))$.

This statement accounts for the influence of factors $q_m, \ldots, q_1$ but does not take into account the influence of interfering factors of the set i^F. When the univocal dependence k_0 is subject to an influence of interfering factors of the set i^F, then sometimes this dependence holds (when the influences of those factors cancel each other) and sometimes not (when the influences of some factors prevail the influences of the remaining ones). The less essential factors belong to i^F, the higher is the percentage of the cases of the first kind when the dependence k_0 occurs, and the reverse. Hence, accounting for the influence of interfering factors results in establishing a probabilistic statement:

(PST^0) the probability of
$F(x) = k_0(B_1(x), \ldots, B_r(x), q_m(x), \ldots, q_1(x))$
on the condition that
$U(x)$ and $\bigvee i_x^F \neq 0$ and $q_1(x) \neq 0$ and ... and $q_m(x) \neq 0$
equals p

where p comes 'sufficiently close' to 1, i.e., $p = 1 - \xi$, ξ being established as the limit of admissible fluctuations. This statement, which will be written as follows:

(PST^0) $Prob\{[F(x) = k_0(B_1(x), \ldots, B_r(x), q_m(x), \ldots, q_1(x))]/U(x)$ and $\bigvee i_x^F \neq 0$ and $q_1(x) \neq 0$ and ... and $q_m(x) \neq 0\} = p$

contains the removed semi-idealizing condition, that is the propositional function

(3) $\bigvee i_x^F \neq 0$

which is read: at least some factors of the set i^F take on values different than zero for an object x. This is why the statement PST^0, being a factual one, is called a *statistical concretization* of the semi-idealizational statement ST^0.

(b) Simple Semi-idealizational Theory

Let us call the sequence of statements

$$ST^m, ST^{m-1}, \ldots, ST^1, ST^0, PST^0$$

where ST^m is a semi-idealizational law and PST^0 is a statistical concretization of ST^0, a simple semi-idealizational theory of the factor F. In the extreme case, when $m = 0$, i.e., when the sequence consists of the semi-idealizational law of the second kind ST^0 and its statistical concretization, we shall call it a semi-idealizational theory of the second kind. When $m \neq 0$, the sequence will be said a semi-idealizational theory of the first kind.

Notice that what is usually called by econometricians an econometric model may be interpreted as a semi-idealizational theory of the second kind. Here is the typical form of the model as presented by econometricians themselves:

(4) $$Y = f(X_1, \ldots, X_k) + \mathcal{N}$$

where Y is a "dependent variable", $X_1, \ldots, X_k$ are "independent variables" and $\mathcal{N}$ is

random component of the model presenting the joint effect of influence upon Y all these factors which are not taken into account as explaining variables within the model (Pawłowski (1), p. 38).

This random component is necessary because within economic life

causal connections occur which are not relations of pure form but their way of manifestation is disturbed by adventitious effects of non-essential, accidental factors. . . . Obviously, no one is able to include all these factors into a model and only the most important variables are treated as explanatory. The result of the joint effect of the remaining factors is taken into consideration only globally through a random component $\mathcal{N}$. . . $\mathcal{N}$ is equal to the difference between the value of the variable Y and the function of explaining variables $X_1, \ldots, X_k$, accepted within the model. In this connection it is often said that the random component is equal to the deviation of the actual value of the variable Y from its theoretical value appointed by the values of the explaining variables (Pawłowski (1), pp. 38–39).

Equation (4) is understood by econometricians as a probabilistic statement, since they state that according to (4) the variable Y has the mean value $f(X_1, \ldots, X_k)$, while in particular cases values of Y deviate from this mean value on the amount of $\mathcal{N}$. The symbol $D(\mathcal{N})$ denotes the variance of the random component:

. . . the stronger the influence of factors that are not taken into account in the model, the greater $D(\mathcal{N})$ . . . In the extreme case, if it was the case they $D(\mathcal{N}) = 0$, we would have a determinist model, that is one which would not contain accidental deviations (*ibid.*, p. 39).

Note that the thesis (4) presupposes a semi-idealizational statement of the second kind equipped with a semi-idealizing condition abstracting from all the essential factors for Y besides $X_1, \ldots, X_k$. Moreover, the latter statement is determinist in character since $D(\mathcal{N}) = 0$, whereas after removing this semi-idealizing condition interfering factors are taken into account and the probabilistic statement (4) is established. One may then observe that the structure of an econometric model of the factor Y is the following:

(5) if $U(x)$ and $\bigwedge i_x^Y = 0$, then $Y(x) = f(X_1(x), \ldots, X_k(x))$,
(6) if $U(x)$ and $\bigvee i_x^Y \neq 0$, then $Y(x) = f(X_1(x), \ldots, X_k(x)) + \mathcal{N}$,

where (6) is a statistical concretization (see econometricians' interpretation of the role of $\mathcal{N}$ quoted above) of the semi-idealizational statement (5).

(c) Simple Semi-Approximation Theory

Another type of theories containing semi-idealizational statements is obtained when concretization of a semi-idealizational statement is not final but ends with approximation. In other words, after ST^m is concretized and $ST^{m-1}, \ldots, ST^j$ are established the procedure of approximation is applied resulting in the following statement:

(AST^j) if $U(x)$ and $\bigwedge i_x^F = 0$ and $q_1(x) \leq a_1$ and ... and $q_{j-1}(x) \leq a_{j-1}$ and $q_j(x) \neq 0$ and ... and $q_m(x) \neq 0$, then $F(x) \approx_\varepsilon f_j(B_1(x), \ldots, B_r(x), q_m(x), \ldots, q_j(x))$

And the statistical concretization of AST^j will take the following form:

($PAST^j$) $Prob\ \{[F(x) \approx_\varepsilon f_j(B_1(x), \ldots, B_r(x), q_m(x), \ldots, q_j(x))]/(U(x)$ and $\bigvee i_x^F \neq 0$ and $q_1(x) \leq a_1$ and... and $q_{j-1}(x) \leq a_{j-1}$ and $q_j(x) \neq 0$ and ... and $q_m(x) \neq a_m)\} = p$.

The sequence of statements

$ST^m, ST^{m-1}, \ldots, ST^j, AST^j, PAST^j$

is called a *semi-approximation simple theory* of the factor F.

(d) Complex Semi-Theories

A *complex semi-idealizational theory* of the factor F is termed the following sequence of systems:

$$(SZ^m, ST^m)$$
$$(SZ^{m-1}, ST^{m-1})$$
$$\dots\dots\dots\dots\dots$$
$$(SZ^1, ST^1)$$
$$(SZ^0, ST^0)$$
$$(PSZ^0, PST^0)$$

where successive theoretical models satisfy the following conditions: (1) SZ^i is the set of semi-idealizational statements logically independent of each other and based on the same i idealizing assumptions; ST^i is a semi-idealizational statement, which follows from the set SZ^i and describes the factor F; (2) SZ^m is the set of semi-idealizational laws; (3) SZ^i is the union of a non-empty set R^i and a set N^i; (4) for every statement t in R^i there is a statement t' in R^{i+1} such that t is its concretization; (5) N^i contains these statements which describe the dependence of those secondary factors which are taken into consideration after the removal of the $i + 1$ – *the* idealizing assumption; (6) the set PSZ^0 contains exclusively statistical concretizations of the statements from SZ^0, and for each statement in SZ^0 there is its statistical concretization in PSZ^0; PST^0 is a consequence of the set PSZ^0 describing the magnitude F ($i = 0, 1, \dots, m$).

And the last form of a scientific theory that may be built within model IV: a *complex semi-approximation theory* of the factor F is called the sequence of systems

$$(SZ^m, ST^m)$$
$$(SZ^{m-1}, ST^{m-1}),$$
$$\dots\dots\dots\dots$$
$$(SZ^j, ST^j)$$
$$(ASZ^j, AST^j)$$
$$(PASZ^j, PAST^j)$$

where theoretical models (SZ^m, ST^m), …, (SZ^j, ST^j) satisfy conditions (1)–(5) of the above definition, and moreover (7) ASZ^j contains exclusively approximations of statements which are in SZ^j and for every statement from the latter set there is its approximation in ASZ^j, (8) $PASZ^j$ contains all and only statistical concretizations of the semi-approximation statements from the set ASZ^j.

(d) Verification of Semi-Idealizational Statements

The difference between verification of idealizational and of semi-idealizational statements consists in applying approximation in the first and statistical concretization in the second case. Hence, the way of verifying the idealizational law runs from T^m to AT^m, while the way of verifying of the semi-idealizational law is the following: from ST^m to AST^m and to $PAST^m$. The latter statement is considered to be confirmed in research practice only if it gives admissible predictions. Here is the definition of the notion of admissible prediction widespread in econometrics: a theory "gives admissible predictions, if for previously appointed numbers $\varepsilon < 0$ and $0 < \gamma < 1$ this condition is satisfied:

$$Prob\{|Y_T - y_{Tp}| < \varepsilon\} \geq \gamma$$

(Pawłowski (2), p. 33), where Y_T is the theoretical value of the variable Y for the time T, y_{Tp} is the actual value of the variable Y for T. And here is the interpretation of this definition within the conceptual apparatus of model IV:

(7) $U(a)$ and $\bigwedge i^Y_a = 0$ and $q_1(a) = 0$ and... and $q_m(a) = 0$ and $Y(a) = Y_T$

(8) $U(a)$ and $\bigwedge i^Y_a = 0$ and $q_1(a) \leq b_1$ and... and $q_m(a) \leq b_m$ and $Y(a) \approx_\alpha Y_T$

where $b_j \leq a_j$ (a_j is the number occurring in an assumption of the general sentence AT^m),

(9) $Prob[Y(a) \approx_\alpha Y_T/(U(a)$ and $\bigvee i^Y_a \neq 0$ and $q_1(a) \leq b_1$ and ... and $q_m(a) \leq b_m)] - \beta$.

The prediction (9) is to be admissible if numbers α, β are not greater than the thresholds of approximation and statistical concretization correspondingly. In order for this prediction to be admissible it is necessary to give it a value of the magnitude foreseen which is 'sufficiently close' to the theoretical value Y_T of the magnitude Y and at the same time such a value the appearance of which 'sufficiently probable.'

It may be observed that the concept of verification changes then within model IV. First of all, the notion of acceptable prediction does not refer to the strict determination of a foreseen value (see models I and II) or even to the approximate determination of the value (see model III), but to the probable approximate determination of it. Second, the Marxian principle of idealizational empiricism (see Chapter 6; and Chapter 10, Section 3) is concretized within model IV. It may be formulated thus: if the statistical concretization $PAST^j$ of the approximation AST^j of the semi-idealizational statement ST^j gives inadmissible (in the above sense) predictions, then the statement ST^j should be rejected, except the case when the statistical concretization $PAST^t$ of the approximation $AST^{t<j}$ of the semi-idealizational statement ST^t (t) gives predictions which turn out to be admissible. Let us present the form of the statements in question:

(ST^j) if $U(x)$ and $\bigwedge i_x^F = 0$ and $q_1(x) = 0$ and...and $q_{j-1}(x) = 0$ and $q_j(x) \neq 0$ and...and $q_m(x) \neq 0$, then $F(x) = f_j(B_1(x), \ldots, B_r(x), q_m(x), \ldots, q_j(x))$

(AST^j) if $U(x)$ and $\bigwedge i_x^F = 0$ and $q_1(x) \leq a_1$ and...and $q_{j-1}(x) \leq a_{j-1}$ and $q_j(x) \neq 0$ and...and $q_m(x) \neq 0$, then $F(x) \approx_\varepsilon f_j(B_1(x), \ldots, B_r(x), q_m(x), \ldots, q_j(x))$

($PAST^j$) $Prob\{F(x) \approx_\varepsilon f_j(B_1(x), \ldots, B_r(x), q_m(x), \ldots, q_j(x))/[U(x)$ and $\bigvee i_x^F \neq 0$ and $q_1(x) \leq a_1$ and...and $q_{j-1}(x) \leq a_{j-1}$ and $q_j(x) \neq 0$ and...and $q_m(x) \neq 0]\} = p$

Let us assume now that the factual statement $PAST^j$ gives inadmissible predictions, i.e., an observational fact appears to be negating this statement. It is supposed to negate also AST^j and ST^j but only on the condition that this fact really refutes the statement ST^j, i.e., that no concretization of ST^j does improve the situation. But let us show that there is such a concretization ST^t of ST^j.

(ST^t) if $U(x)$ and $\bigwedge i_x^F = 0$ and $q_1(x) = 0$ and...and $q_{t-1}(x) = 0$ and $q_t(x) \neq 0$ and...and $q_m(x) \neq 0$, then $F(x) = f_t(B_1(x), \ldots, B_r(x), q_m(x), \ldots, q_t(x))$

which is approximated:

(AST^t) if $U(x)$ and $\bigwedge i_x^F = 0$ and $q_1(x) \leq a_1$ and...and $q_{t-1}(x) \leq a_{t-1}$ and $q_t(x) \neq 0$ and...and $q_m(x) \neq 0$, then $F(x) \approx_{\varepsilon'} f_t(B_1(x), \ldots, B_r(x), q_m(x), \ldots, q_t(x))$

and statistically concretized:

($PAST^t$) $Prob\{F(x) \approx_{\varepsilon'} f_t(B_1(x), \ldots, B_r(x), q_m(x), \ldots, q_t(x))/ [U(x)$ and $\bigvee i_x^F \neq 0$ and $q_1(x) \leq a_1$ and...and $q_{t-1}(x) \leq a_{t-1}$ and $q_t(x) \neq 0$ and...and $q_m(x) \neq 0]\} = p'$

where $\varepsilon' < \varepsilon$, since accounting for some additional factors allows for decreasing the deviations with experience, and $p' \geq p$, since taking into consideration these additional determinants may raise the probability of realization of the respective dependence. In this case our concretization of the Marxian principle of idealizational empiricism forces the rejection of $PAST^j$ and AST^j but not of ST^j; the latter is preserved as it has been shown that the respective fact only apparently negates ST^j.

3. IDEALIZATION AND DETERMINISM

(a) Criteria of a Deterministic Nature of Theories.

Let us distinguish two parts of a scientific theory: a *factual system* consisting of the last model of the theory, i.e., containing exclusively factual statements, and a *system of idealizations* consisting of all the previous models, i.e., containing idealizational or semi-idealizational statements (the latter of a strict or approximate form). Some criteria of the deterministic nature of a theory may be now discussed.

The first criterion would state that any theory which satisfies the two conditions: (i) its system of idealizations consists of univocal statements only and (ii) its factual system, consisting exclusively of univocal statements which are concretizations of idealizational laws drawn from the system of idealizations, is deterministic. This criterion would imply that there are theories in which corrections with respect to all disturbing factors, which cause deviations from the relationships described in those theories, are rigorously taken into account. This, however, seems improbable as the classic example of a deterministic theory, that of Newtonian mechanics, is not deterministic in the sense specified above, since that theory certainly has its 'approximation part'.

The second criterion would state that a theory is deterministic if and only if it has (i) a system of idealizations that consists of univocal statements only, (ii) a factual system that consists exclusively of univocal concretizations or approximations of laws drawn from its system of idealizations. In the light of this criterion, appropriate methodological analyses would be needed to decide which scientific theories are, and which are not, deterministic. At any rate it may be said that this criterion is much more liberal then the former. In particular, in the latter case a deterministic theory is not expected to formulate univocal predictions as to the future states of objects. Such predictions are formulated with a margin of toleration, which is required not for such trifling reasons as accidental errors in measurement, but for the essential reason that not all those corrections which should be made in view of the disturbing factors are rigorously taken into account; they are replaced by a 'summary correction" which indicates the admissible level of deviations jointly for all those factors which are left out in the process of strict concretization.

Finally, the third criterion would be as follows: a theory is deterministic if and only if its system of idealizations consists exclusively of univocal statements. In this case, the factual system may consist of both concretizations and univocal approximations or probabilistic concretizations. Under this criterion, a deterministic theory might yield not only approximate predictions, but also probabilistic predictions of the last-named kind. The essential point is only whether the regularities described by the system of idealizations of a given theory are univocal or statistical. This criterion is thus still more liberal than the second. When Einstein made his by now famous statement to Born about God playing dice, he perhaps had in mind a criterion that would come close to the third one, as he was obviously concerned with the nature of the regularities investigated by science, and not with precision in predictions.

(b) The Causal Nexus

But if we agree to distinguish between idealizational and approximational laws and if we accept the fact that practically every idealizational law – next to a possible strict concretization – is subject to an approximation and statistical concretization, following which its approximate analogue is established, then we have to reformulate the problem of the causal interpretation of laws of science. This is so because the question arises whether causal relationships state semi-idealizational laws that are in the system of idealizations of a given theory, or their statistical concretizations which are in the factual system of that theory. Both cases will be analysed separately, starting with the second.

The fact that it would be paradoxical to say that formulas stating causal relationships hold under idealizing conditions, seems to speak in favour of the second possibility. The relation of causality is often defined by reference to such concepts as that of pressure, which apparently pertain to relations between physical objects. Likewise, examples of causal relations current in philosophical works are always associated with such events as a billiard ball striking another billiard ball, the rubbing of a match against the match-box and the burning of the match, etc. This may, however, turn out to be less paradoxical if we realize that in mechanics we consider such facts as the collision of two point-masses and that reference is often made in this connection to a causal relation. Moreover, even the simplest examples of causal relationships drawn from everyday life reveal that the analyses carried out by philosophers are based on tacit idealizational assumptions. It is claimed, for instance, that the causal relation is transitive. But consider three billiard balls, *A*, *B*, and *C*, in their actual conditions, such as the friction of the ball against the surface of the billiard table, the resistance of the air, etc. Now it is obvious that the striking of ball *A* with a cue which results in *A* striking *B* need not result in *B* striking *C*, if only the disturbing factors are sufficiently strong (Naganowski (1)). It is, however, tacitly assumed that the disturbing factors are eliminated. But if this is so, then, by tacitly adopting a counterfactual assumption, we in fact analyse the movements of 'ideal billiard balls', and not those of real balls (at least not directly the movements of real balls).

Consider now the first possibility. Its adoption results in the need to distinguish between an 'ideal' causal nexus and a real causal trend. The former is stated in the appropriate semi-idealizational law, the latter, in a factual statement which is an approximation and statistical concretization of that semi-idealizational law. The causal nexus holds between colliding point-masses, and the causal trend between colliding physical bodies. The causal nexus is exception-free, whereas the causal trend holds in the majority of known cases of a specified kind. The causal nexus is manifested in the causal trend in the sense that the occurrence of specified events in the physical world depends, first, on the occurrence of the factor described in the semi-idealizational law ('the main cause'), and, secondly, on the occurrence of those factors ('side-causes') which are disregarded in the semi-idealizational law on the strength of appropriate counterfactual assumptions, but which do occur in fact and hence 'disturb' the working of the main factor. It is because of these disturbing factors (secondary or interfering ones) that the causal nexus does not manifest itself in fact (is not manifested in its 'pure form'), but is manifested as the causal trend. It is, perhaps, such ideas – which would, of course, require a more precise formulation – that are latent in the well-known Marxian thesis that a regularity manifests itself by random events:

In theory it is assumed that the laws of capitalist production operate in their pure form. In reality there exists only approximation, but this approximation is the greater, the more developed the capitalist mode of production and the less it is adulterated with survivals of former economic conditions (Marx (1), Vol. III, p. 172).

Regularity always manifests itself

as a tendency, i.e., as a law whose absolute action is checked, retarded, and weakened, by counteracting circumstances (*ibid.*, p. 229).

(c) Is a Statistical Explanation Possible?

According to the Marxian model of explanation in order to explain a fact it is necessary to refer to the appropriate idealizational law (Chapter 8, Section 4(a)). This requirement was maintained within models II and III. Now it has to be modified because of the introduction of a new type of counterfactual statements, viz. semi-idealizational ones. When the influence of interfering factors upon the magnitudes under investigation is allowed within our model of scientific activity, the above-mentioned requirement of the essential nature of explanation has to be reformulated. It says presently that in order to explain a fact it is necessary to refer to the appropriate semi-idealizational law. Therefore the modification of the Marxian model of explanation which is proceeded in model IV is this:

$$ST^{m} \dashv ST^{m-1} \dashv \ldots \dashv ST^{j} \rightsquigarrow\!\dashv AST^{j} \dashv_{s} PAST^{j} \wedge P \rightarrow E$$

where $\dashv_{s}$ stands for the relation of statistical concretization. What is accordingly necessary for explaining is to start from a semi-idealizational law revealing what is treated as an essence of a given type of fact and abstracting not only from factors considered to be secondary ones but also from factors the investigator thinks to be interfering for the magnitude in question.

An important consequence follows from this: since semi-idealizational laws are univocal statements, then an explanation which starts from probabilistic statements is not allowed at all. That is, probabilistic statements may be used in explanation but only as intermediate links of explanatory chains – they may not initiate any explanatory chain. Therefore probabilistic explanations are not correct explanations in the light of the presented theory of science. It seems to me that this conclusion results from ideas that may be found in the works of the classics of Marxism. Since "the general law acts as the prevailing tendency only in a very complicated and approximate manner, as a never ascertainable average of ceaseless fluctuations" (Marx (1), Vol. III, p. 159), then the statement describing this 'general law' may be interpreted as a semi-idealizational statement which has to be statistically concretized. Therefore statements of a statistical nature are allowed to occur as the latest links of explanatory chains only: "It is this law that explains the deviations, and not vice versa, the deviations that explain the law" (Marx (1), Vol. III, p. 184).

CHAPTER 12

PROGRAMMING AND PRACTICAL SCIENCES[1] (MODEL V)

1. EXPLAINING AND PROGRAMMING

(a) Removal of p_1

Our ideal researcher from model IV is much more similar to actual investigators than its more idealized counterpart from model I: he applies the method of deduction in developing a theory, resorts to approximations, uses statistical statements, and so on. But he also aims at explanation of observed facts as his primary goal. This is so because of assumption p_1 being in force. But there is a significant part of scientific activity that does not satisfy this condition. For the so-called practical sciences (engineering, agricultural sciences, medical sciences, etc.) not explanation but programming of future facts seems to be peculiar. Therefore, we remove now this condition which makes possible the analysing of peculiarities of practical research.

(b) An Analysis of the Example

Let us start with discussing an example of programming which shows the way economists solve the problem of the optimum rate of investments in gross national product (GNP)[2].

I (A) The problem of determining the optimum rate of investments in GNP is being discussed with the help of some simplifying assumptions, for example:

(q_1) production property does not depreciate;

(q_2) the growth of property results instantly in the growth of GNP (i.e., the retardations of GNP at the growth of production property are equal to zero);

(q_3) a constant percentage of GNP is allocated at investments each year;

etc.

(B) In Domar-Harrod's theory the following equations describe the growth of an economy which satisfies the above conditions:

(I.1) $i_t = \mathrm{d}m_t/\mathrm{d}t$,
(I.2) $i_t = sy_t$,
(I.3) $\mathrm{d}m_t/\mathrm{d}t = k(\mathrm{d}y_t/\mathrm{d}t)$,

where i denotes production investments, t the time, s the percentage of GNP allocated at investments, m the production property, y the size of GNP, and k the capital-absorbency coefficient.

(C) From equations (I.1)–(I.3) it follows that

(I.4) $(i/y_t)(\mathrm{d}y_t/\mathrm{d}t) = s/k.$

The following formula is the solution of the above equation:

(I.5) $y_t = y_0 \exp(st/k)$

where y_0 is the size of GNP at the time treated as the starting point).

(D) The task consists in determining the rate of investments in GNP which will make consumption maximal in the period $t_0 - t_1$. The problem is thus reduced to finding such s, for which:

(f) $$C = \int_{t_0}^{t_1} (1 - s)y\mathrm{d}t = max$$

whereas

$$y_t = y_0 \exp(st/k).$$

Moreover, some limiting conditions are accepted:

(w_1) consumption does not fall below some norms: $s_{min} \leq s$.
(w_2) the growth of consumption lasts some definite time: t_0 to t_1;

etc.

(E) The problem is solved by a calculation of C finding s at which the integral reaches its maximum. It turns out that the following formula states the condition for it:

(I.6) $$s(1 - s) = \frac{k}{t_1}(1 - \exp(-st_1/k))$$

This solution is insufficiently realistic. For example, if $t_1/k > 2$ for $t \to \infty$, then the rate of investments grows rapidly to the value one.

This would imply that the rate of investments in an economy which maximizes consumption should be extremely high. However, experience gained from economic practice indicates quite the opposite. This discrepancy between the model and reality follows from highly idealized nature of it.

II (A) Let us, therefore, reject assumption (q_1) replacing it with the realist claim that:

(r_1) the production property r_t that requires for replacement in t is greater than zero,

whereas conditions (q_2)–(q_3) are still valid.

(B) Under the above assumptions the economic growth is described by the following equations:

(II.1) $i_t - r_t = \mathrm{d}m_t/\mathrm{d}t,$
(II.2) $r_t = i_{t-v},$
(II.3) $i_t = sy_t,$
(II.4) $\mathrm{d}m_t/\mathrm{d}t = k(\mathrm{d}y_t/\mathrm{d}t),$

where r_t denotes the production property which requires replacement in t, and v the period of exploitation of the production property.

Some of the above equations pass from model I to model II without changes, some other are modified. Moreover, a new equation (II.2) is added; it tells us what a newly introduced secondary factor depends on.

(C) The following formula for the growth of GNP may be derived from (II.1)–(II.4):

(II.5) $y_t = y_0 \exp(wt)$

where w is the rate of the growth of GNP:

$$w = \frac{s}{k}\,(1 - \exp(-v)).$$

(D) The task consists in finding s, for which:

(f) $$C = \int_{t_0}^{t_1} (1 - s)y\mathrm{d}t = max$$

whereas:

$$y_t = y_0 \exp(wt)$$

and conditions (w_1), (w_2) are met. The goal-function is the same as within model I, but one of the conditions has changed–the one which is a conclusion from the accepted theory of economic growth.

(E) One calculates the integral C and finds s at which the integral reaches its maximum. The condition for this is the following:

$$\text{(II.6)} \qquad s(1-s) = \frac{k}{(1-\exp(-v))t_1}\{1 - \exp[-(1-\exp(-v))t_1]/k\}$$

This solution is already more realistic than the previous one – the time v of exploiting the production property is accounted for within model II. Nonetheless, the investment policy in real economy still cannot be based on it. Model II is still the so-called single factor model – it silently assumes that the size of GNP depends on production property alone. Hence it is necessary to concretize model II in turn in order to take into account, for instance, the labour force and other factors GNP depends on.

2. A RECONSTRUCTION OF PRACTICAL RESEARCH

(a) General Scheme of Practical Research

Let us try to generalize the example discussed above. A *criterion of efficiency* will be called the formula:

$$(E) \qquad V = g(F_0, F_1, \ldots, F_n) = \text{extremum}$$

where V is a magnitude being at the same time a social value (see Chapter 14, Section 3), F_0 is a magnitude which can be manipulated practically, that is, one which depends, at least within some range, on somebody's actions, and $F_1, \ldots, F_n$ are other magnitudes which may be practically manipulable, though not necessarily so; $n \geq 0$.

Let us notice that goal-criterion (f) falls under the scheme of a criterion of efficiency: it is a valuable thing in our culture to make GNP as large as possible and s is a practically manipulable magnitude; besides, in this case $n = 1$.

The assumptions of the criterion of efficiency are twofold. On the one hand, practical branches of science are needed in society to determine optimum ways of realizing the social values. Thus, the applied sciences always presuppose some pool of values. On the other hand, it is necessary for an axiological problem (what is the best way to accomplish a given value?) to be scientifically solved to assume some knowledge of the value-factor in question; it should be known what other factors it depends on.

This knowledge may be a conclusion from some theory or a purely empirical generalization. In both cases the following statement is to be accepted:

(b–E) $V = g(F_0, F_1, \ldots, F_n)$

which will be termed the *basis of the criterion* (*E*).

Starting with the definite assumed knowledge and the pool of values accepted in a given society, the practical investigator formulates the criterion of efficiency together with some assumptions. On the one hand, he does so introducing some limiting conditions, that is, formulas of the following form:

$$(W)\qquad \begin{array}{ll} F_{i_1} \leq a_{i_1} & (\text{or } a_{i_1} \leq F_{i_1}) \\ \cdots\cdots\cdots & \cdots\cdots\cdots \\ F_{i_r} \leq a_{i_r} & (\text{or } a_{i_r} \leq F_{i_r}) \end{array}$$

where $i = 0, 1, \ldots, n$, and $i_r \leq n$. The conditions (w_1)–(w_3) fall under the scheme (*W*). The role of limiting conditions is to determine admissible solutions of the problem of extremizing the criterion of efficiency.

On the other hand, our 'practitioner-investigator' looks for a theory which would tell him what the arguments of the magnitude-value *V* depend on. Let us assume the approximation theory of the factors $F_0, \ldots, F_n$, i.e., a theory in the sense of model III (see Chapter 10, Section 1)

(I) $(Z^m, T_0^m, \ldots, T_n^m)$
(II) $(Z^{m-1}, T_0^{m-1}, \ldots, T_n^{m-1})$
....
(*N*) $(Z^i, T_0^i, \ldots, T_n^i)$
$(N+1)$ $(AZ^i, AT_0^i, \ldots, AT_n^i)$

assuming also that the factors $F_0, \ldots, F_n$ described in the statements $T_0^j, \ldots, T_n^j$ have only one principal determinant. Otherwise our scheme would be too complicated. Therefore, the statement T_g^j is of the following form:

(T_g^j) if $U(x)$ and $q_1(x) = 0$ and ... and $q_{j-1}(x) = 0$ and $q_j(x) \neq 0$ and ... and $q_m(x) \neq 0$, then $F_g(x) = k_j(B_g(x), q_m(x), \ldots, q_j(x))$

$(g = 0, 1, \ldots, n, j = m, m-1, \ldots, i)$. To simplify the notation even more we shall omit the individual variable x and the realistic condition $U(x)$.

Next, the problem of the optimum solution is posed within the theory accepted in the first model:

(*Z–O*.I) find F_0, such that:

(a) $V = \alpha(F_0, F_1, \ldots, F_n) =$ *extremum* if at the same time:

(b) $F_{i_1} \leq a_{i_1}$ (or: $a_{i_1} \geq F_{i_1}$)

....

$F_{i_r} \leq a_{i_r}$ (or: $a_{i_r} \geq F_{i_r}$)

(c) if $q_1 = 0$ and ... and $q_m = 0$, then $F_g = k_g(B_g)$ (for $i = 0, 1, \ldots, n, g = 0, 1, \ldots, n, i_r \leq n$).

From the conjunction (a) and (b) and (c) the *first* (*starting*) *optimum solution* or the *optimum law* is being deduced. It is of the following form:

(*R–O*.I) if $q_1 = 0$ and ... and $q_m = 0$, then $F_0 = m(B_1, \ldots, B_n, a_{i_1}, \ldots, a_{i_r})$

and hence

if $q_1 = 0$ and ... and $q_m = 0$, then $V = \alpha(m(B_1, \ldots, B_n, a_{i_1}, \ldots, a_{i_r}), F_1, \ldots, F_n) =$ *extremum.*

The starting solution is valid under m idealizing conditions and, as a rule, it is too abstract. Therefore the assumption $q_m = 0$ is removed and the problem of optimum solution is posed on the basis of the second model of accepted theory:

(*Z–O*.II) find F_0, such that:

(a) $V = \alpha(F_0, F_1, \ldots, F_n) =$ *extremum* if at the same time:

(b) $F_{i_1} \leq a_{i_1}$ (or $a_{i_1} \leq F_{i_1}$)

....

$F_{i_r} \leq a_{i_r}$ (or $a_{i_r} \leq F_{i_r}$)

(c) if $q_1 = 0$ and ... and $q_{m-1} = 0$ and $q_m \neq 0$, then $F_g = k_{g-1}(B_g, q_m)$; $i = 0, 1, \ldots, n, i_r \leq n, g = 0, 1, \ldots, n$.

From the conjunction (a), (b) and (c) the second optimum solution can be drawn (the first concretization of the optimalizational law). It assumes the following form:

(*R–O*.II) if $q_1 = 0$ and ... and $q_{m-1} = 0$ and $q_m \neq 0$, then $F_0 = s(B_1, \ldots, B_n, a_{i_1}, \ldots, a_{i_r}, w(q_m))$

hence:

if $q_1 = 0$ and ... and $q_{m-1} = 0$ and $q_m \neq 0$,
then $V = \alpha[s(B_1, \ldots, B_n, a_{i_1}, \ldots, a_{i_r}, q_m), F_1, \ldots, F_n] = extremum$.

Note that our example may be interpreted now in terms of the above construction. The equations of economic growth (I.1)–(I.3) preceded by the idealizing assumptions (q_1)–(q_3) fall under the set Z^m while their consequence, (I.5), is a particular case of the thesis T^m. From the criterion of efficiency (f), the thesis (I.5), and the limiting conditions, (w_1)–(w_2), the statement (I.6) is deduced, which defines maximal consumption in a given period (it falls under the scheme (*R–O*.I)). The removal of the assumption (q_1) follows and the reference to the second theoretical model included in the theory of growth we are dealing with. The theses (II.1)–(II.4) preceded by the assumptions (r_1), (q_2), (q_3) fall under the set Z^{n-1}, and their consequence, (II.5), under the thesis T^{n-1}. From the criterion of efficiency (*f*), the thesis (II.5) and the limiting conditions, the thesis (II.6) is being deduced, which states that consumption will be maximal in a given period. It falls under the scheme (*R–O*.II).

One proceeds in an analogous way by removing the idealizing assumptions: 'q_{m-1}', '$q_{m-2} = 0$', ..., $q_i = 0$'. The optimum problems are raised: '*Z–O*.III, *Z–O*.IV, ...'. They refer to the concretizations $T_0^{m-2}, \ldots, T_n^{m-2}$ and $T_0^{m-3}, \ldots, T_n^{m-3}$, etc. The problems are solved by deducing the optimum solutions (*R–O*.III), (*R–O*.IV), ... from the premisses (a), (b) and $T_0^{m-2}, \ldots, T_n^{m-2}$, and then from the premisses (a), (b), and $T_0^{m-3}, \ldots, T_n^{m-3}$, etc. The final results are that we account for the factors $q_m, \ldots, q_i$ and obtain $q - i + 1$ optimum solution. Sometimes the solution is a sufficient approximation for making a decision.

The sequence of systems:

$$((a), (b), T_0^m, \ldots, T_n^m; (R\text{–}O.\text{I}))$$
$$((a), (b), T_0^{m-1}, \ldots, T_n^{m-1}; (R\text{–}O.\text{II}))$$
....
$$((a), (b), T_0^i, \ldots, T_n^i; (R\text{–}O.N))$$
$$((a), (b), T_0^0, \ldots, T_n^0; (R\text{–}O.N + 1))$$

where the solutions of theoretical models I–*N* of the accepted theory are linked with the relation of concretization, and the solutions of the model $N + 1$ are the approximations of the solutions of the model *N*, is called the *optimum model*. In constructing the optimum model, an investigator employs both the solutions of successive theoretical models of the theory he had accepted, and the premisses which express his mode of evaluation (or, more precisely, the mode of evaluation which is expected of him because of his role of a practical researcher).

Then a practical investigator designs a *real model* (a prototype of a machine, organizational model of an economy, etc.) so that it meets the final solution (R–$O.N + 1$) while meeting the conditions (a), (b) and $T^0_0, \ldots, T^0_n$. The real model is subsequently tested in order to find out whether the intentions of the maker are realized, i.e. whether it does, in fact, meet the optimum solution with a sufficient approximation (R–$O.N$); whether it meets the final optimum solution (R–$O.N + 1$). The construction of a real model and its evaluation can be performed by different

The Scheme of Practical Research

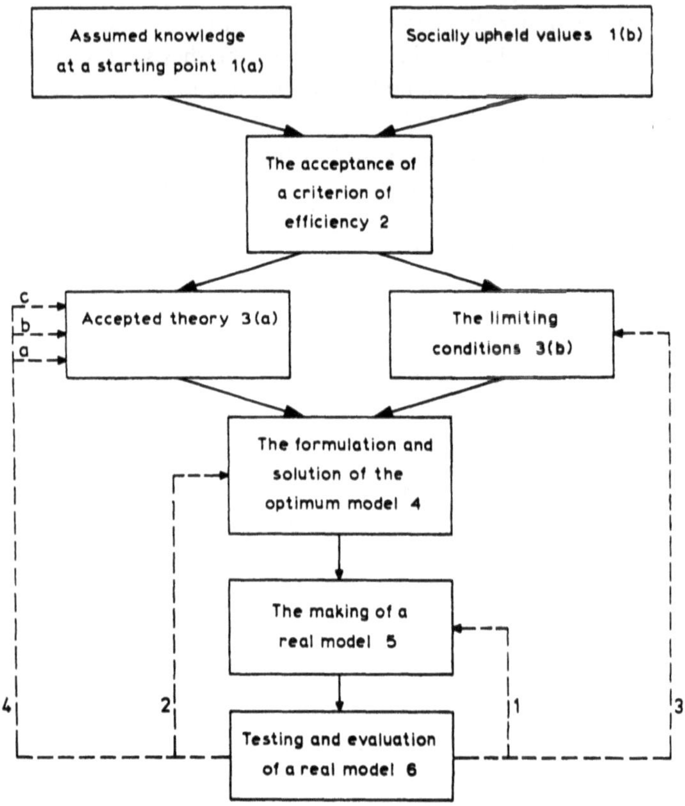

Fig. 6

persons, and that is sometimes connected with the division of social rates. We aggregate these roles for the sake of simplicity and speak about a practical investigator.

If a discrepancy between the characteristics of the real model appears, the validity of earlier operations is being controlled. First, one controls the accuracy of the execution of a real model with respect to the optimum model. If a mistake is discovered, the correction and another test and evaluation follow. If the criterion of efficiency or the limiting conditions are not met in spite of the corrections, the second stage of control follows – a revision of the optimum model by means of modification of the accepted theory. There are three types of modification:

(a) further concretization of the statements from Z^i follows; a real model is then produced on the basis of approximation of the solution given by one of more realistic models; this model is evaluated, etc., as above;

(b) if the discrepancies are still significant, the investigator concludes that the very theory (physical or chemical for instance) is apparently incomplete and additional factors influencing the investigated phenomena have to be discovered and explicitly abstracted from which will enable him to concretize a new form of a theory with respect to their influence;

(c) a repetition of further stages follows; if the results are not compatible with the criterion of efficiency or the limiting conditions, then the practical investigator has to construct an appropriate theory by himself (that is, he is forced to play the role of a theoretician) or revises (usually, weaknes) the limiting conditions; the criterion of efficiency is never revised – it may happen that at a given stage of knowledge the criterion in question is unattainable.

The scheme of practical investigation can finally be presented as in Figure 6.

(b) Practical Versus Theoretical Investigation

The basic difference between theoretical and practical investigation consists in the latter having a peculiar criterion of theorems control. In theoretical branches of science an accordance with experience is basically the criterion of acceptance. In practical sciences such a criterion would lack justification since it is known *a priori* that optimum solutions are not satisfied in the world of human acts and their results. If they were, in fact no practical research would be needed. For it is their task to the program of achieving the optimum investment rate, a possibly efficient machine, etc., a quite different criterion of control is applied in these sciences. It is the criterion of accordance with the previously accepted formula of efficiency (E), if the postulated limiting conditions are satisfied. Optimum models are rejected in practical sciences not because 'things are different in the world', but since they do not satisfy sufficiently social evaluations (imposed on a researcher by society).

The above difference follows, however, from a more fundamental difference between practical and theoretical investigations. The principal role of the latter is to explain phenomena, while the principal role of the former is to programm them, that is, to describe them so that they could be optimal (i.e., so that the criterion of efficiency could be satisfied under limiting conditions). The procedure of programming can be described as composed of the following sequence of operations: the deduction of the first optimum solution from the criterion of efficiency, limiting conditions and the solutions of the principal model of the accepted theory; the deduction of the second optimum solution from the criterion of efficiency, limiting conditions and the concretizations of the solutions of the principal model, etc. It may be said that the procedure of programming consists in the procedure of optimum model-construction.

(c) A Conjecture Concerning the Status of Normative Ethics

Though the matter requires a separate study, the traditional problem of the status of the so-called axiological sciences (normative ethics, normative aesthetics) may be attempted to solve on the basis of the presented conception.

It seems that 'axiological research' can be treated as a special form of a practical investigation. Nobody has ever asked, for instance, 'how to be happy?' but assumed, at least tacitly, that some limiting conditions are satisfied – that no other 'true people' will be hurt, that it will not be achieved at excessive cost, etc. Also, a psychological or sociological theory was, at least tacitly, accepted. It is well known that the concept of 'true people' evoluted, and so did the concept of happiness. But it seems that the methodological scheme of considerations did not change and it was, and is, similar to the model of practical research outlined above. According to the latter, some criterion of efficiency and a theory are assumed, some limiting conditions are applied, and, from all of these, conclusions are derived referring to the possible realization of an ideal determined by the criterion of efficiency. To be sure, there are some differences between 'axiological discourse' and 'practical discourse' but it seems that they are rather of a secondary importance being a result of backwardness of the former.

The conjecture formulated above, that 'axiological knowledge' is a retarded 'practical knowledge', has two important consequences. First, it follows that, contrary to the commonly held views, the 'axiological discourse' does not possess a peculiar methodology but it falls under the same methodology that all practical sciences have[3]. Second, assuming the conjecture in question one may explain why the problems of, let us say, normative ethics are still so controversial as they are. They are controversial thanks to the same factor due to which technical problems were controversial before the rise of modern science – namely, thanks to the lack of a theory enabling to solve optimum problems more precisely. It is not particularly difficult to ask for values. What is really difficult is to determine the conditions under which these values would be best satisfied, since we lack theories necessary for optimization of a vast number of values. It is not so difficult to ask: 'what should be the angle of the gun barrel, if the gun is to fire the longest possible distance?'. The difficulty consists in determining this angle – it was Galileo's theory which was necessary to solve that problem.

3. FURTHER CONCRETIZATIONS

(a) General Remarks

Our methodological models I–V are of increasing complexity and, I hope, of accuracy in explaining what is really done in empirical sciences. Model I presents the basic (from the viewpoint of my interpretation of the Marxian ideas) features of scientific activity: the idea of a law as the 'deepest' (usually idealizational) statement, the idea of a theory as organized by the relation of concretization of idealizational statements, the idea of explanation as referring to the 'deep structure' revealed in the law and developed in its successive concretizations, whereas models II–IV modify these features. The structure of a scientific theory becomes more and more complicated: it is additionally determined – within model II by the relation of entailment, within model III by the relation of approximation, within model IV by the relation of statistical concretization. At the same time, some (in models II–III) or all (in model IV) of the remaining basic features of scientific activity shown in model I are appropriately modified. In model V nothing is changed within theoretical scientific activity, but it is complemented by the image of practical scientific activity and the interrelations between them. To sum up: our models become more and more realistic images of scientific activity.

In spite of this, even model V presents scientific activity in a very oversimplified manner. It is due to the fact that it still presupposes some additional idealizing conditions abstracting from additional features of this activity. Let us briefly discuss some of them. Our starting point will be model I which makes it possible to show most clearly the essence of the modification in question.

(b) Variations of Idealization

A closer inspection of our examples reveals that examples taken from actual research practice do not always fall strictly under proposed idealized methodological schemes. It is not surprising that this is the case. Our models are concerned with some selected aspects of scientific enquiry and abstract from the remaining ones. Therefore these models may be referred to actual scientific practice only approximately (i.e., by approximation). This is the case for all idealizational theories in every branch of science. Nonetheless, some of these discrepancies between the idealized pattern of science and science itself can be reduced thanks to taking into account some previously disregarded aspects of scientific enquiry.

Let us consider, for instance, assumption p_7 of the Marxian law of value (Chapter 1, Section 1):

(1) $S(x) - D(x) = 0$

saying that the difference between the supply of a commodity x and the demand for this commodity equals zero. Strictly speaking, this assumptions, being undoubtedly very close to idealizing ones with respect to its form and function, does not come under the definition of an idealizing condition. According to this definition a propositional function:

(2) $p(x) = 0$

is of an idealizing character if there is no actual object possessing the property p in the degree 0. Given this definition one may state that the function

(3) the dimensions of x equal zero

is an idealizing one, while the function (1) is not, since there is not excluded an existence of a commodity (although this would be a rather rare case) being at a given moment in conditions of market equilibrium. Shortly, the existence of a material point is excluded whereas the existence of market equilibrium for a given commodity is not.

Now, one may notice that our methodological construction, even in its most developed form in model V, does not explain the role of conditions like (1). Even their existence is excluded due to accepted definitions. This proves that all the models in order to be acceptable are to be equipped with an additional assumption which leads to elimination of the conditions under consideration. The assumption would say that adventitious factors are at work (i.e., take on values other than zero) whenever the principal factors for a given model are at work; hence there is no situation, according to this idealizing assumption, in which a magnitude under consideration F is subject to the working of principal factors alone. This condition should be joined to the list of conditions p_1–p_5 of model I obtaining in this way model I′ which is more abstract than model I. After removal of p_5 model II′ is obtained which differs from model II only by being more abstract due to the assumption in question. The same is valid for models III′, IV′ and V′. The latter contains only one idealizing assumption, that which has been recently formulated.

Let us remove this assumption. Thus it is admitted that principal factors are at work when only some secondary factors or even none of them act. Having such situations admitted we have to take into account a new type of counterfactual conditions. Let U be the universe of discourse. The condition:

(4) $p(x) = 0$

is called a *quasi-idealizing* assumption, if there is such a proper subset K of the universe U that for every $a \in K$: $p(a) = 0$ and for every b in the remaining part of U (i.e., for each $b \in U - K$): $p(b) \neq 0$. The set K is the *range of realization* of the condition (4), while the set $U - K$ is the *range of idealization* of this condition. For instance, (1) is a quasi-idealizing condition, since some commodities may (which, it is true, does not occur frequently) be in a state of a market equilibrium, while for the others this is not the case.

A universal statement which in its antecedent has realistic, and idealizing conditions only will be termed an *idealizational statement of the first kind*; if it is equipped with realistic, idealizing and quasi-idealizing conditions, then it is termed an *idealizational statement of the second kind*; and if the repertoire of the assumptions is limited to realistic and quasi-idealizing ones, then such a statement is an *idealizational statement of the third kind*. The first type of idealizational statements corresponds to what has been called simply idealizational statements (in models I–V). We can accordingly single out idealizational laws of the first, the second and the third type: they include at any rate all those assumptions which seem to be necessary to eliminate those secondary factors which do not co-occur with the principal factors. Therefore, a universal conditional is a law of science within model I′, is a scientific laws of the first, second, or third kind).

Quasi-idealizing conditions in science are removed in a somewhat different way than idealizing assumptions. This can best be shown in the case of an idealizational statement of the third kind:

(5) if $U(x)$ and $q(x) = 0$, then $F(x) = k(B_1(x), \ldots, B_r(x))$.

Let the range of realization of the quasi-idealizing condition '$q(x) = 0$' be denoted by K, while the range of idealization of this condition is denoted as $U–K$. Let also the following two factual statements be true:

(6) $Z(x)$ iff $x \in K$,
(7) $Z'(x)$ iff $x \in U–K$.

Now the removal of the quasi-idealizing assumption under consideration yields two statements:

(8) if $U(x)$ and $Z(x)$, then $F(x) = k(B_1(x), \ldots, B_r(x))$
(9) if $U(x)$ and $Z'(x)$, then $F(x) = s(k(B_1(x), \ldots, B_r(x)), t(q(x)))$.

Statement (8) is a limited concretization of (5), while (9) is a degenerate limited concretization of (5). When a quasi-idealizing assumption is removed, a given statement is modified only as far as its scope of idealiza-

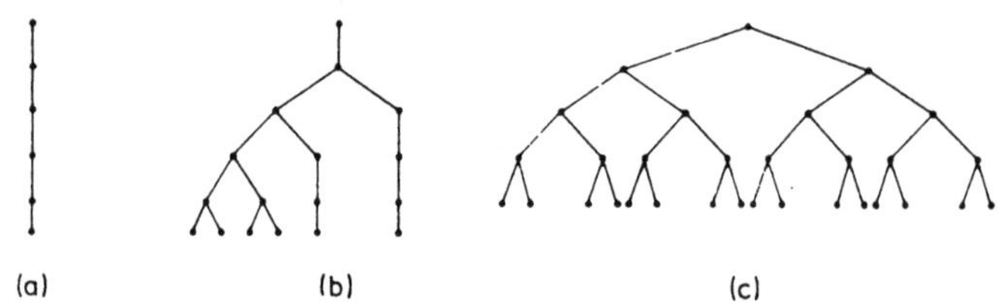

Fig. 7. (a) The structure of an idealizational theory of the first kind. (b) The structure of an idealizational theory of the second kind: single lines denote the relation of concretization, while pairs of lines (<) denote the relation of limited concretization (one of the subsequent statements is a limited concretization and the other is a degenerated concretization of the initial statement). (c) The structure of an idealizational theory of the third kind: each statement is connected to the pair of statements as its (limited or degenerated) concretizations.

tion is concerned; when it comes to the scope of realization the said statement remains unchanged. Concretization in the previous sense may be called a strict one.

As a result, the structure of a simple idealizational theory from model I is replaced within model I′ with the structure of line (*idealizational theory of the first kind*) or with the structure of a thin tree (*idealizational theory of the second kind*) or with the structure of a thick tree (*idealizational theory of the third kind*) (see Figure 7):

If assumption p_5 of our theory was removed, then model II′ would be constructed and complex idealizational theories of the first, second and third kind have to be distinguished. In model III′ being a result of the removal of conditions p_4 and p_3 one has to define the relation of limited approximation and to distinguish complex approximation theories of the three types. Etc.

In this way finding out new idealizing conditions which have to be accepted if our theory is to be able to explain scientific facts leads to

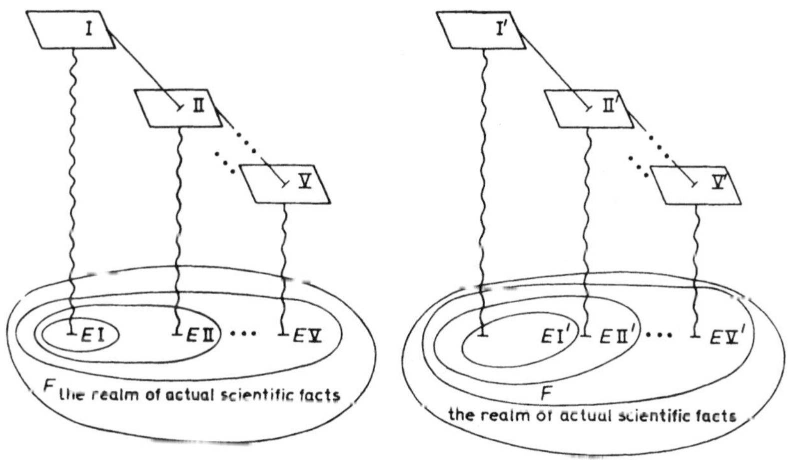

Fig. 8. (a) The fact F falsifies the theory consisting of models I–V; it does not belong to the scope of facts explained by this theory (i.e., to EV). (b) The fact F confirms the modified theory consisting of models I′–V′; it does belong to the enlarged scope of facts EV′ explained by the new version of this theory. The scope of explanation of a model of a theory is linked with this model by the relation of approximate explanation (i.e., explanation in the ordinary sense including the approximation of this model) marked with ⤳, while models themselves are connected to each other with the relation of concretization ⊣ .

modifications of the theory. This modified form of the theory (its concretization) explains those facts which falsify the initial theory. In our case, occurrence of conditions of the type (1) falsified the theory consisting of models I–V, since their role was not understandable in the light of models I–V and even some examples shown as confirming model I (the law of value – Chapter 1, Section 1) were not, strictly speaking, confirmation for this model. It has been said that the law of value is an idealizational law in the sense of model I (which was maintained in models II–V) whereas, speaking precisely, this was not the case; the law of value as containing at least one quasi-idealizing condition (see above) is not an idealizational law of the first but of the second kind. At the same time, this very instance which falsified models I–V does not do this after removal of the new idealizing assumption, i.e., after establishing model I′. Since the law of value which is not, strictly speaking, an idealizational statement in the sense of model I, *is* an idealizational statement in the sense of model I′, i.e., it is an idealizational theorem of the second type. What falsifies model I does confirm model I′. This situation may be shown graphically as in Figure 8.

(c) The Meta-Methodological Remark

Also the modified version of our theory which has been briefly outlined above has to be based on some additional idealizing conditions if it is to retain its validity. Many such conditions have been shown and the appropriate corrections have been made. Brzeziński *et al.* (1) have shown that the theory qresupposes that factors are one-sidedly essential for each other, while it is quite known that factors sometimes mutually influence each other. The modification proposed by the authors allows for explanation as to why empirical theories contain *many* axioms in their initial models – since the set of axioms (i.e., basic idealizational laws called assumptions in our terminology) has to reveal the *mutual* connections of factors. In the same paper another modification has been proposed with respect to the assumption saying that each factor influences a given magnitude at most in one way only. Since there occur factors determining some other magnitudes in many ways within different scopes of objects, then theories describing such factors should also be included into the idealizational concept of science which has been done in this work where the structure of the theories of this kind has been analyzed. As Patryas (1) has shown, further concretization of the theory in question may be done which allows an explanation of the role of *ceteris paribus* clause which in turn yields a significant change in the so far accepted concepts concerning verification of laws (see Patryas (2)). Brzeziński (3) and Zielińska (1) have analyzed corrections connected to the mutual influences within a given space of essential factors. Zielińska (2) has proved that accounting for the matters connected with units of the magnitudes in question allows for a considerable enlargement of the scope of statements falling under the schemes of an idealizational law and of a concretization of it. Brzeziński (2) and Tuchańska (3) have considered corrections of the idealizational theory of verification with respect to the problem of operationalization of approximation statements; the simplified version of the theory presupposes that all approximation statements (or respective statistical concretizations) are equipped with definite criteria of empirical applicability. Tuchańska (2) has accounted for the problems of construction of images of factors revealing that the idealizational concept of theory is based on the tacit assumption that an investigator entirely faithfully recognizes objective factors although this is evidently not the case in actual scientific practice. Many other works of this type exist which reveal tacit assumptions of the idealizational theory of scientific activity and correct this theory (Patryas (3), Nowak (8) and (9), Brzeziński (4), Kozówna (1), Stefański (1), Tuchańska (1), Chwalisz (1), Nowakowa (1), Kocikowski (1), Banaszak (1), and others).

What is perhaps more important now is that the idealizational theory of scientific activity has a peculiar methodological feature which seems to me to be worthy of noting. A characteristic property of almost every contemporary methodological trend is the incompatibility between the image of science it presents and the image of itself it proposes. According to inductivism, science has to observe and generalize observations, while no one of the inductivists seriously claims that this is what a methodologist does. According to instrumentalism, empirical science constructs a theory as pure calculus which does not say anything about the world being simply an intermediate link between observables, while no instrumentalists would say that the thesis of instrumentalism is a 'pure formula' which says nothing about the structure of science or that Hempel's model of explanation does not say anything about actual explanations taking place in empirical sciences. Finally, according to hypotheticism, empirical hypotheses apply the method of criticism of hypotheses with the aid of empirical tests, while as the author of *The Logic of Scientific Discovery* has put it, methodology determines 'conventions', concerning scientific conduct. In a word: inductivism is not a product of induction, instrumentalism does not have only instrumental value, the conception of hypotheticism is not itself a result of applying the method of criticism of hypotheses. The idealizational methodology, however, can be interpreted, roughly, as a kind of idealizational theory (i.e., this theory may be approximately treated as falling under our methodological schemes). Therefore this methodology employs the same method which is employed by the sciences with which it is concerned.

NOTES

[1]This chapter is based on the article Chwalisz *et al.* (1).
[2]The substantial data are contained in Czerwiński (1).
[3]This is Znamierowski's (1) idea.

CHAPTER 13

SCIENTIFIC COMMUNITY AND PROGRESS OF SCIENCE

1. THE STARTING POINT OF THE THEORY OF COGNITIVE PRACTICE

In several previous chapters I have tried to develop an idealizational theory of scientific activity, or, in other words, an idealizational methodology. Our object of interest was a cognitive subject whose aim is to understand phenomena and who employs definite procedures, with that of idealization as the principal one, to achieve this goal. This image was slightly modified in model V when practical sciences were taken into account, but theoretical sciences remained a standard pattern of scientific activity. In this way, due to isomorphism of a simple idealizational theory and of essential and nomological structures the methodological theory presented in Chapter 8, Section 2(d) realizes the idea of science as the reflection of reality (Chapter 7, Section 3(c)). This has been done within the presented methodological theory.

There is, however, another idea concerning science among ideas traditionally accepted in Marxism – that cognitive progress is a kind of social progress. This thought cannot be expressed within our methodological theory which is concerned with cognitive actions of a typical investigator, that is, of a single knowing subject (Chapter 8). A theory of cognitive practice is, then, necessary to explicate and develop the idea in question. The starting point of this theory is, however, the methodological theory of scientific activity. Only assuming the characteristics of a particular investigator which this theory gives us, we may establish some principles governing the development of the body of science conceived as a joint result of a scientific community. And only assuming such principles which tell us, in particular, what is the progress of science are we able to show connections between science and social practice. In this sense, the idea of science as the reflection of reality has a priority in comparison with the idea of cognitive process as a kind of social process: in order to explain the latter the former has to be presupposed, but the reverse does not hold.

2. PROGRESS OF SCIENCE

(a) Theoretical Orientation

Each idealizational theory is based on a definite ontological perspective, limiting an image of the essential structure of a given magnitude to a definite kind of images. Different idealizational theories may assume the same ontological perspective and they may also assume different perspectives of the type. A scientific theory based on a definite ontological perspective is termed a *theoretical orientation*. The latter can be, then, identified to a system (P, T) where P stands for a definite ontological perspective, while T is a *scientific theory*. For the sake of simplicity we shall discuss the most basic case where T is a simple idealizational theory of a given factor. Two theoretical orientations may differ with respect to the first or to the second component. A scientific trend will be called the set of theoretical orientations accepting the same ontological perspective. Theoretical orientations which belong to the same trend propose different theories, e.g. referring to different factors as determinants of a given magnitude, but all the factors will be of the same type or referring to different dependencies but these will be dependencies of the same kind.

(b) The Social Nature of the Growth of Science

Let us assume that a theoretical orientation is composed of a theory T of a factor F, while another one of a theory T' of the same factor. The former is thus the system

$$O = (P, T),$$

while the latter is the system

$$O' = (P', T').$$

P may be identical with P' but we do not presuppose this. Thus the followers of these two orientations can belong either to the same scientific trend or to different ones. Let us also assume that the following image of the essential structure of F is accepted within O:

$$H,$$
$$H, p.$$

Thus the theory T, in the simplest case, is composed of two statements:

T^1: if $U(x)$ and $p(x) = 0$, then $F(x) = f(H(x))$
T^0: if $U(x)$ and $p(x) \neq 0$, then $F(x) = g(f(H(x)), h(p(x)))$.

Let us also assume that within O' another image is assumed for the same magnitude F:

L,
L, p.

Then the theory T' is composed of, let us say, of two statements:

S^1: if $U(x)$ and $p(x) = 0$, then $F(x) = k(L(x))$
S^0 if $U(x)$ and $p(x) \neq 0$, then $F(x) = n(k(L(x)), h(p(x)))$

Of course, the statement T^0 demonstrates some discrepancies if compared with the results of experiments:

(1) $U(a)$ and $p(a) \neq 0$ and $F(a) = m$

where

(2) $m \neq g(f(H(a)), h(p(a)))$

but

(3) $|m - g(f(H(a)), h(p(a)))| \leq \varepsilon$;

ε is the threshold of approximation accepted within a given science. If the theory T' satisfies analogical conditions, then the followers of a theoretical orientation O begin to view the new orientation O' as a serious partner. For our ideal scientists it is not the reason for eclectic addition to the statements of the theory T those of T', but it is the real stimulus for correcting their own theory T.

Namely, within the older theoretical school there is a growing conviction that the factor L considered to be the principal one in the light of orientation O' is indeed influencing the factor F. And since it has been neglected within O, they include it as a secondary for F, that is, they enlarge the image of the essential structure of F within their own orientation. A new image is built:

H,
H, L,
H, L, p.

Hence a new variant of the theory T is proposed within the older school, one composed of the following statements:

T^2: if $U(x)$ and $p(x) = 0$ and $L(x) = 0$, then $F(x) = f(H(x))$

T'^1: if $U(x)$ and $p(x) = 0$ and $L(x) \neq 0$, then $F(x) = s(f(H(x)), k(L(x)))$

T'^0: if $U(x)$ and $p(x) \neq 0$ and $L(x) \neq 0$, then $F(x) = u[s(f(H(x)), k(L(x))), h(p(x))]$.

If the factor F is actually essential for F, then the statement T'^0 gives us a better approximation to the empirical data than the older thesis T^1.

The same process can occur within the rival orientation O'. Here also a construction of the new variant of the theory T' composed of the theses S^2, S'^1, S'^0 takes place; their form may be easily shown. Here, too, the process of correction results in a better agreement with empirical data, if the factor H is actually essential for F. Victory belongs to the rival who achieves the highest degree of agreement with empirical data for the largest range of facts.

It should be added that the simplified image of competition between some theoretical orientations acquires some practical importance when orientations from different scientific trends occur as partners, i.e., when they have different ontological perspectives. Each theoretical orientation can grow autonomously – using empirical discoveries of new essential factors of the types distinguished by its ontological perspective. However, factors of different types than the distinguished one go, as a rule, unnoticed within the given perspective; it is not merely an accident that Marxism did not notice psychological factors influencing people's behaviour and psychologism in sociology did not distinguish economic mechanisms of the social growth. If the factors of the types provided by another ontological perspective are indeed essential for investigated phenomena, they can be taken into account only by means of polemics against this orientation of a quite different type than our own. But this polemics result in a significant improvement of our own theory. That is the reason why we should have as many opponents as possible. The principle of tolerance follows from our cognitive interest.

(c) Approaching the Absolute Truth

A *relative truth* is a statement which represents a regularity or its non-surface form of manifestation, whereas an absolute truth is a statement representing the surface form of the regularity (Chapter 8, Section 6 (e)). The image of the essential structure of the respective magnitude assumed by the appropriate absolute truth must be identical to the essential structure itself. On the other hand, the image assumed by the relative truth has only to be basically similar to this structure (i.e., the factors considered to be the ones have to be actually principal for the magnitude in question). Therefore the above mentioned mechanism of 'completing' the images of the essential structure as the result of competition between the theoretical orientations leads through the sequence of relative truths toward the absolute truth only in case when at the starting point a relative truth has been achieved, i.e., when respective principal factors have been discovered. Let us consider this more carefully. What, however, should be repeated before is that this mechanism of scientific growth is of a social nature because the social nature of doing science makes it possible.

3. MODELS OF SCIENTIFIC GROWTH

(a) Assumptions

The theory of cognitive practice in undoubtedly the much less developed part of the idealizational concept of science than the theory of scientific activities. Therefore it is only the latter which can be presented in the form of an idealizational theory. In case of the former only partial attempts of doing this are possible. At present I shall try to construct such a fragmentary theory of the growth of science realizing the ideas presented above.

Let us start with terminological stipulations. Since investigators can never be quite sure whether they have correctly recognized the essential structure, some pragmatic notions will be useful. Let us call a relative (absolute) statement the one which is believed to be a relative (absolute) truth. Now idealizing conditions of our partial theory may be introduced. The basic, most abstract of our models will be equipped with the following assumptions

(Z. 1) The repertory of those factors which are believed to be principal for a given magnitude is not revised in the successive stages of the development of knowledge.

(Z. 2) The repertory of these factors which are assumed to be secondary for a given magnitude is not revised in the successive stages of the development of knowledge.

(Z. 3) A researcher working in the ith period in the development of science knows how the ith secondary factor affects the magnitude under consideration.

Now, (Z. 3) assumes that we treat the history of science as divided into a number of periods; even if the researcher who formulates a given idealizational statement does not know how a secondary factor at the ith level of significance works, then, in accordance with (Z. 3), a researcher working in the $(i + 1)$th period (stage) in the development of knowledge (science) will know that.

(b) Model I

Consider now what a rationally acting researcher (and we are concerned with such researchers only) will do if he intends to explain all facts in their complexity and also knows how a given secondary factor modifies the idealizational statement he has formulated. Obviously, he will concretize his statement. Hence, if (Z. 1) to (Z. 3) are assumed, the development of science follows the schema shown in Figure 9.

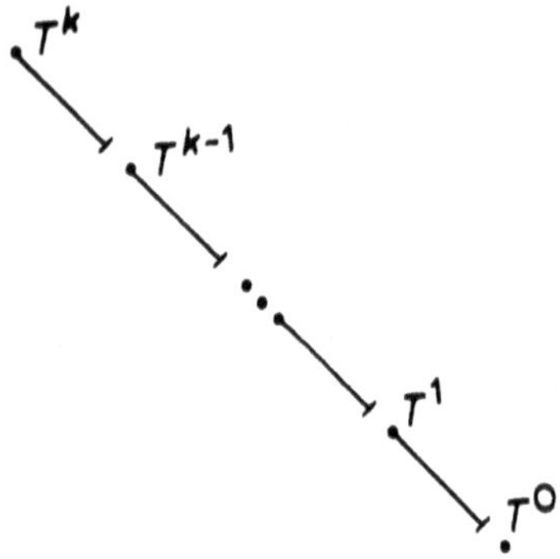

Fig. 9. Within Model I scientific growth is defined by the relation of concretization (⊣).

The superscript i indicates that a given statement has i idealizing assumptions. Hence, if a statement T^k is in the form:

T^k: if $U(x)$ and $p_1(x) = 0$ and ... and $p_k(x) = 0$ then $F(x) = f_k(H(x))$.

then in the next period it is concretized so as to yield:

T^{k-1}: if $U(x)$ and $p_1(x) = 0$ and ... and $p_{k-1}(x) = 0$ and $p_k(x) \neq 0$, then $F(x) = g_1[f_k(H(x)), h_1(p_k(x))]$.

In the next period the concretization procedure is applied further, until in the kth period we arrive at the factual statement:

T^0: $U(x)$ and $p_1(x) \neq 0$ and ... and $p_k(x) \neq 0$,
then $F(x) = g_{k-1}[f_1(H(x), ..., p_2(x)), h_{k-1}(p_1(x))]$.

We see that in Model I it is true that the correspondence sequence is a sequence of increasingly complete relative statements, and the last term in that sequence is an absolute statement.

By a *correspondence sequence* we mean just a sequence of successive statements which follow one another in the course of the development of science and which refer to one and the same magnitude. A statement t is said to be *more complete* than a relative statement t' if and only if t is a concretization of t' The formulation 'more complete' is justified by the fact that t represents a fuller determination of one and the same magnitude as it takes into account certain secondary factors which are disregarded in t'.

We see that in Model I an absolute truth is, theoretically, attainable. For if a researcher has established a proper image of the essential structure for a given magnitude, and if $T^k, ..., T^0$ are true statements, then $T^k, ..., T^1$ are relative truths ordered by the increasing degrees of completeness, and T^0 is an absolute truth. If (Z.1) (Z.3) are assumed, then absolute truth is attainable under the conditions specified above. This would comply with the intuitive formulations often to be found in works by Marxist philosophers:

> Progress in science consists in the transition from one relative truth to another relative truth . . . . and the ideal goal in the form of an absolutely precise description of given facts ought to be termed an absolute truth (Krajewski (1), p. 216).

Thus, in Model I, the development of science is based on the relation of strict concretization. But this is so only in Model I, under the idealized conditions defined by the assumptions (Z.1) – (Z.3).

(c) Model II

We now remove (Z.3) and thus assume realistically that it sometimes occurs that a researcher does not know, at a given moment, how a given secondary factor affects a given magnitude, and he accordingly rests satisfied with an approximation. The last-formulated realistic assumption now replaces the idealizing condition (Z.3). In Model II we thus have the idealizing conditions (Z.1) and (Z.2) and the realistic condition in question.

Now things stand thus: a researcher turns his initial idealizational statements into concretizations as far as he can, say up to the statement

T^i, and then approximates that statement. After a period of time one of his continuators acquires some knowledge about the ways in which the secondary factor p_i affects the magnitude in question, and he accordingly removes the idealizing assumption '$p_i(x) = 0$' by resorting to the concretization procedure. Other researchers later arrive at the idealizational statement T^f, which they are unable to concretize as they do not know how the factor P_f works, and have to rest satisfied with an approximation of that statement, etc. Hence, in Model II, the schema of the development of science is shown in Figure 10.

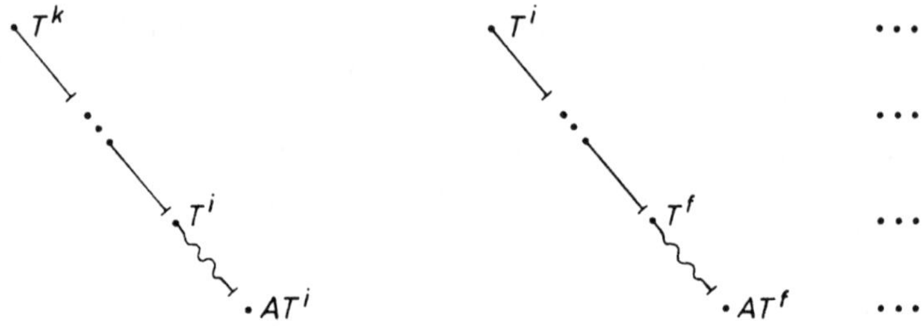

Fig. 10. Within Model II scientific growth is defined by the relation of concretization (⊣) and of approximation (~⊣).

Hence, in Model II, it is true that the correspondence sequence includes relative statements T^k, ..., T^i, ..., T^f of increasing completeness, and the approximate statements AT^i, AT^f, ..., which are in that sequence, are of increasing precision.

To prove this it suffices to refer to the concepts of concretization and approximation. The strict concretization of the idealizational statement T^k yields:

T^i: if $U(x)$ and $p_1(x) = 0$ and ... and $p_i(x) = 0$ and $p_{i+1}(x) \neq 0$ and ... and $p_k(x) \neq 0$, then $F(x) = g_{k-i+1}[f_{k-i+2}(H(x), p_k(x), \ldots, p_{i+2}(x)), h_{k-i+1}(p_{i+1}(x))]$.

Its approximation yields the following factual statement:

AT^i: if $U(x)$ and $p_1(x) \approx 0$ and ... and $p_i(x) \approx 0$ and $p_{i+1}(x) \neq 0$ and ... and $p_k(x) \neq 0$, then $F(x) \approx g_{k-i+1}[f_{k-i+2}(H(x), p_k(x), \ldots, p_{i+2}(x)), h_{k-i+1}(p_{i+1}(x))]$.

At a later period, T^i is subject to strict concretization and yields the idealizational statement

T^f: if $U(x)$ and $p_1(x) = 0$ and ... and $p_f(x) = 0$ and $p_{f+1}(x) \neq 0$ and ... $p_i(x) \neq 0$ and ... and $p_k(x) \neq 0$, then $F(x) = g_{k-f+1}[f_{k-f+2}(H(x), p_k(x), \ldots, p_{f+2}(x)), h_{k-f+1}(p_{f+1}(x))]$.

The statement T^f is next subjected to approximation:

AT^f: if $U(x)$ and $p_1(x) \approx 0$ and ... and $p_f(x) \approx 0$ and $p_{f+1}(x) \neq 0$ and $p_i(x), \neq 0$ and ... and $p_k(x) \neq 0$, then $F(x) \approx g_{k-f+1}[f_{k-f+2}(H(x), p_k(x), \ldots, p_{f+2}(x)), h_{k-f+1}(p_{f+1}(x))]$.

If we now compare AT^i with AT^f we see immediately that AT^f is more precise than AT^i as it takes into account the additional magnitudes $(p_i, \ldots, p_{f+1})$, which are disregarded in AT^i.

Note that it follows immediately from this that, once the researcher has established the proper image of the essential structure of a given magnitude, then $T^k, \ldots, T^i, \ldots, T^f$ are relative truths (of increasing completeness), and the sequence AT^i, AT^f, ... is one of increasingly precise *approximate truths*. Thus, in Model II. the absolute truth cannot be attained, but is being approached by a sequence of approximate truths, the successive terms of that sequence yielding results of increasing precision; those results accordingly differ less and less from those which the absolute would yield if it could be formulated effectively. In Model II the progress of science consists in an incessant approach to the absolute truth which, however, need not be attained. On the other hand, the assumptions made in Model I predetermined its attainment at one moment, be it even extremely remote.

(d) Model IIIA

The assumptions made in the models discussed so far excluded any modification of the list of secondary factors, once these have been assumed. But, of course, such modifications do take place in the development of science. As a rule, we learn at a later period that certain additional secondary factors, whose existence originally was not supposed, are at work. This realistic condition will be made to replace (Z.2). Thus, the set of assumptions in Model IIIA consists of the idealizing assumption (Z.1) and the realistic assumptions replacing (Z.2) and (Z.3).

In order to avoid a symbolism that would become excessively complicated we will use a simpler example to explain the development of science as shown by Model IIIA. Let the initial significance structure of a magnitude F be:

H
H, p_3
H, p_3, p_2
$H, p_3, p_2, p_1.$

A researcher convinced that this significance structure is a proper one, establishes the basic idealizational statement:

T^3: if $U(x)$ and $p_1(x) = 0$ and $p_2(x) = 0$ and $p_3(x) = 0$, then $F(x) = f_3(H(x))$.

and then subjects it to strict concretization, say, in two stages:

T^2: if $U(x)$ and $p_1(x) = 0$ and $p_2(x) = 0$ and $p_3(x) \neq 0$, then $F(x) = f_2(H(x), p_3(x))$,
T^1: if $U(x)$ and $p_1(x) = 0$ and $p_2(x) \neq 0$ and $p_3(x) \neq 0$, then $F(x) = f_1(H(x), p_3(x), p_2(x))$.

Suppose now that they do not know how the secondary factor p_1 affects F and hence resort to the approximation:

AT^1: if $U(x)$ and $p_1(x) \approx 0$ and $p_2(x) \neq 0$ and $p_3(x) \neq 0$, then $F(x) \approx f_1(H(x), p_3(x), p_2(x))$.

Suppose, however, that a new factor q, which is significant for F, comes to be found at a later period. Suppose also that q is as significant as p_2. Hence the new significance structure, which is now assumed to be proper, is in the form:

$H,$
$H, p_3,$
$H, p_3, p_2,$
$H, p_3, p_2, q,$
$H, p_3, p_2, q, p_1.$

A researcher accordingly may not now rest satisfied with T^3 because he knows that that statement is based on idealizing assumtions which are too weak. He has to disregard the factor q. Hence he joins that additional idealizing assumption to T^3 which then becomes

T'^3: if $U(x)$ and $p_1(x) = 0$ and $q(x) = 0$ and $p_2(x) = 0$ and $p_3(x) = 0$, then $F(x) = f_3(H(x))$.

Successive concretizations yield

T'^2: if $U(x)$ and $p_1(x) = 0$ and $q(x) = 0$ and $p_2(x) = 0$ and $p_3(x) \neq 0$, then $F(x) = f_2(H(x), p_3(x))$,

T'^1: if $U(x)$ and $p_1(x) = 0$ and $q(x) = 0$ and $p_2(x) \neq 0$ and $p_3(x) \neq 0$, then $F(x) = f_1(H(x), p_3(x), p_2(x))$.

The statements T'^3 to T'^1 differ from their analogues T^3 to T^1 from the earlier period only by the idealizing assumption which disregards the newly discovered secondary factor q. Now that additional idealizing assumption is removed and T'^1 is subjected to concretization:

T''^1: if $U(x)$ and $p_1(x) = 0$ and $q(x) \neq 0$ and $p_2(x) \neq 0$ and $p_3(x) \neq 0$, then $F(x) = k_1(H(x), p_3(x), p_2(x), q(x))$.

This statement has no analogue among the statements formulated in the earlier period, because at that time the working of the factor q was not taken into account at all. Suppose that the effect of p_1 on F is not known still, so that the researcher this time too must resort to the approximation procedure:

AT''^1: if $U(x)$ and $p_1(x) \approx 0$ and $q(x) \neq 0$ and $p_2(x) \neq 0$ and $p_3(x) \neq 0$, then $F(x) \approx k_1(H(x), p_3(x), p_2(x), q(x))$.

As can be seen, the introduction of the new factor q in a later period has the following consequences: the former statement T^3 must be made more abstract (transition is made in science from T^3 to T'^3) in the sense that it now includes a new idealizing assumption which disregards that factor q; further, if researchers, as has been assumed, strive to explain facts, the said statement must be concretized with respect to q which yields the statement T''^1. In other words, the statement T'^3 *dialectically corresponds* to T^3 in the sense introduced by Nowakowa (1). In Model IIIA, which is more realistic than the two models of the development of science which have been discussed earlier, the development of science is thus determined by the principle of dialectical correspondence, which diagrammatically can be shown as in Figure 11.

As can be seen, in the first period science develops after the schema shown in Model II. Next it is discovered that in fact an additional secondary factor is at work, too. When this is found out, all previous statements $T^k, \ldots, T^i$ must be made more abstract if the said factor turns out to be significant for the magnitude under consideration at the ith level. But then the statement T'^j (obtained from T^i by making the latter more abstract) must be concretized with respect to the working of the factor q.

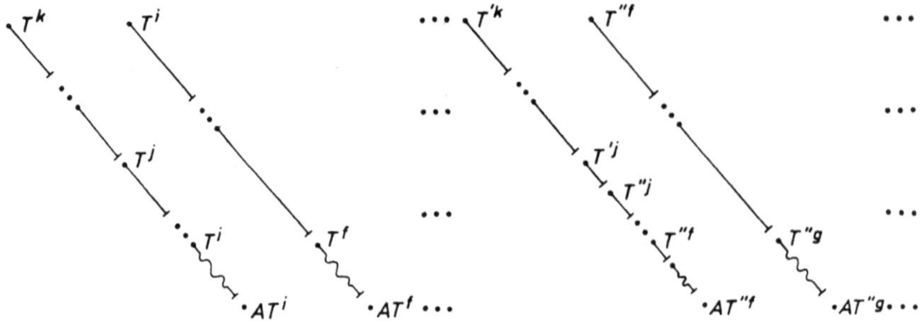

Fig. 11. Within Model IIIA stages of growth are connected with the relation of dialectical correspondence.

This yields the new statement T''^j, which is subjected to further approximation. At that moment the mechanism described by Model II starts working: researchers formulate a sequence of approximate statements of increasing precision. Then, at one time a new secondary factor r whose effects have not been realized so far, is discovered, and so on.

This shows that, in Model IIIA, the development of science follows the principle of dialectical correspondence. Due to this the working of that principle will be illustrated by a brief example. The reconstruction will not be complete, but will go as far only as is necessary to illustrate the concept of dialectical correspondence. In classical mechanics, the following formula for velocity been adopted:

(4) if $B(x)$, then $v(x) = a(x)t(x)$.

where B stands for the set of physical bodies, v, for velocity, a, for acceleration, and t, for time. The special theory of relativity shows that formula (4) must include an additional idealizing assumption, that is, one which is not satisfied in the light of the theory of relativity:

(5) if $B(x)$ and $1/c = 0$, then $v(x) = a(x)t(x)$.

where c stands for the velocity of light. Thus, statement (4) has been made more abstract. On the other hand, it is subjected to concretization:

(6) if $B(x)$ and $1/c = \text{const}$, then $v(x) = \dfrac{a(x)t(x)}{[1 - a^2(x)t^2(x)/c^2]^{1/2}}$.

Should we apply the symbolism adopted previously to the example now under consideration, then (4) would have to be symbolized by T^0, its more abstract version (5) by T'^0, and (6) i.e., the concretization of (5), by T''^0.

When we compare (5) with (4) we see that (5) is more essentially true than (4) as it includes a fuller set of idealizing assumptions. In general terms: if T'^k dialectically corresponds to T^k, then T'^k is more essentially true than T^k.

It may now be said that the following relationship holds in Model IIIA: the correspondence sequence includes relative statements T^k, T'^k . . . of increasing adequacy; their concretizations T^i, T'^i, T^f, T'^f . . . are, respectively, relative statements of increasing completeness; and the approximate statements AT^i, AT^f, ..., AT''^f, AT''^g . . . are marked by increasing precision.

Note that the relationship between T'^k and T^k is such that only the assumptions are changed, whereas the consequent remains the same. Hence, if T^k is satisfied non-vacuously by some objects (theoretical or hypothetical objects, i.e., ideal types) then we can find a set of assumptions which makes it possible to eliminate the falsifying objects and to leave in our field of vision the verifying objects only. In other words, if T^k is satisfied non-vacuously by any objects at all, then in the sequence T^k, T'^k, ... there is a statement, say T'^k, such that if it assumes a proper structure and if T'^k, ..., T'^i, T''^i, ..., T''^f, are essentially true statements, then T'^k, ..., T'^j, ..., T''^f are relative truths, and AT''^f, AT''^g . . . are approximate truths of increasing precision.

Thus Model IIIA also includes a mechanism of coming closer to the absolute truth, even though that mechanism is not as simple as in the earlier models. Yet Model IIIA is still based on the assumption (Z. 1) that the list of principal factors is not subject to any revision.

(e) Model IIIB

Let that assumption now be removed. This will realistically allow for these cases in which the opinion on what is the principal factor for a given magnitude changes in science. The problem will again be illustrated by a simplified example.

Let the significance structure for F have originally the same form as that discussed previously. Then researchers establish the basic statement T^3, its concretizations T^2, T^1, and the approximation AT^1 of the last-named idealizational statement T^1. But then it turns out in a later period that the factor H, which was originally believed to be the principal one for F, is not the principal one but a secondary one, whereas it is the magnitude G which is the principal factor. In other words, in that later period the significant structure is as follows:

$G,$
$G, H,$
$G, H, p_3,$
$G, H, p_3, p_2,$
$G, H, p_3, p_2, p_1.$

Hence in that later period an entirely new original law is formulated in the form:

S^4: if $U(x)$ and $p_1(x) = 0$ and $p_2(x) = 0$ and $p_3(x) = 0$ and $H(x) = 0$, then $F(x) = k_4(G(x))$.

When this law is concretized with respect to the working of the magnitude H, researchers make use of the former statement T^3 which shows how F depends on H:

S^3: $U(x)$ and $p_1(x) = 0$ and $p_2(x) = 0$ and $p_3(x) = 0$ and $H(x) \neq 0$, then $F(x) = g_1[k_4(G(x)), h_1(H(x))] = k_3(G(x), H(x))$.

where $h_1(H) = f_3(H)$.

As can be seen, in that later period use is made of the former knowledge of the working of the factor H, which originally was held to be the principal one, and now is held to be a secondary. The next stage consists in concretizing S^3 with respect to the working of the other secondary factors:

S^2: if $U(x)$ and $p_1(x) = 0$ and $p_2(x) = 0$ and $p_3(x) \neq 0$ and $H(x) \neq 0$, then $F(x) = k_2(G(x), H(x), p_3(x))$.

S^1: if $U(x)$ and $p_1(x) = 0$ and $p_2(x) \neq 0$ and $p_3(x) \neq 0$ and $H(x) \neq 0$, then $F(x) = k_1[G(x), H(x), p_3(x), p_2(x)]$.

Suppose a researcher does not know how F is affected by p_1. He then rests satisfied with the approximation:

AS^1: if $U(x)$ and $p_1(x) \approx 0$ and $p_2(x) \neq 0$ and $p_3(x) \neq 0$ and $H(x) \neq 0$, then $F(x) \approx k_1[G(x), H(x), p_3(x), p_2(x)]$.

The schema of the development of science as reflected by Model IIIB can easily be imagined. It would consist of successive schemata shown in Model IIIA. Each period in which the dialectical correspondence principle operates, as drawn from that model, would begin with new a basic statement. There are certain links which connect the various periods that are governed by the principle of dialectical correspondence. This applies first of all to the fact, mentioned above, that basic statements current in an earlier period are later used to explain the way in which those factors operate which formerly used to be treated as principal and are later believed to be secondary. Nevertheless, that model does not reveal any mechanism which would enable us to say that science somehow approaches the absolute truth.

(f) The Model of 'Permanent Revolution' and the Model of 'The Threshold of Maturity'

The models discussed above have been treated merely as theoretical constructions. The time has now come to find out what relation they bear to empirical data, in our case, to facts drawn from the history of science.

We have to note first that the assumption (Z.1) is somewhat different in nature from the remaining two assumptions on which Model I is based: it is a quasi-idealizing assumption, that is satisfied for some kinds of empirical objects, but is not satisfied for the remaining ones. It sometimes really does occur in science that researchers change their opinions on what is the principal factor for a given magnitude. But it also does occur that the same principal factors remain valid for centuries even though the list of secondary factors is being revised from time to time. It is therefore to be concluded that Model IIIA and Model IIIB may be treated as equally realistic. It is true that they are also based on many other idealizing assumptions which, however, will be disregarded here as they are less important for the issues in the development of science. Suppose, therefore, that both models are sufficiently realistic. What then is the picture of the development of real science they give?

We see here two possible solutions. One is to assume that in real science changes in principal factors are normal phenomena, and revolutions in science consist in changes in the repertory of principal factors. This might be defined as the model of a 'permanent revolution'. The other is to assume that changes in the lists of principal factors are exceptional phenomena which are decisive for the attainment by science of the stage of maturity. Revolutions in mature science would consist in radical changes in secondary factors, important for numerous statements, or in completions (but not revisions) of the list of principal factors; they would thus be governed by the dialectical correspondence principle. This interpretation might be defined as the model of 'the threshold of maturity': before the period of maturity we witness revisions of principal factors in science, but from a certain moment on those principal factors become more or less fixed and their list can at the most be completed, while secondary factors can be revised; from the period of maturity the development of science is governed by the dialectical correspondence principle.

(g) Preliminary Empirical Data

There is no place here, of course, to analyse the history of science from the point of view of the two interpretations. And the choice of one of the two rival approaches must, self-evidently, be based on empirical considerations.

We can nevertheless refer to a number of papers which offer fragmentary reconstructions of the history of selected laws in empirical sciences. These data form a picture of the development of science which seems clearly to confirm the model of "the threshold of maturity'.

For instance, the relationship between Galileo's principle of inertia and the analogous principle to be found in Aristotle's works is not determined by the dialectical correspondence principle, as we one dealing with a revision of the list of principal factors. For Aristotle, they are: external pressure and the resistance given by the milieu; for Galileo, it is inertia. Those factors which were the principal ones for Aristotle are secondary in classical mechanics: they are taken into account in further concretizations (Nowakowa (5)).

But from Galileo on, the dialectical correspondence principle began to govern physics. Newton's law on freely falling bodies dialectically corresponds to Galileo's (Such (1), Nowakowa (3)). Newton's law of gravitation dialectically corresponds to Kepler's laws (Nowakowa (3)). The successive formulations of Ohm's law are governed by the dialectical correspondence principle (Nowakowa (4)). The various theorems in the theory of relativity bear to theorems of classical mechanics the relation of dialectical correspondence, which is linked with the fact that Lorentz' transformations dialectically correspond to those of Galileo (Nowak (10)).

These preliminary empirical data seem to point to the correctness of the model of "the threshold of maturity", but the issue is still largely open and can be settled only by possibly numerous and varied methodological reconstructions of cases drawn from various empirical disciplines.

(h) The Attainability of the Absolute Truth

The fact that those empirical data which are available so far point to the correctness of the model of 'the threshold of maturity' makes us draw an important conclusion. As has been said, Model IIIA includes a mechanism which, under specified conditions, ensures approximation to the absolute truth. On the other hand, Model IIIB does not include such a mechanism. Such being the case, the presumable correctness of the model of 'the threshold of maturity' implies that once that threshold is crossed in science (which in physics apparently took place at the time of Galileo) we have to do with a gradual approach to the absolute truth, which is due to the operation of the mechanism presented in Model IIIA. This might be understood as a confirmation of the traditional Marxist idea that science approaches the absolute truth through a series of relative truths. And, having our interpretation of this mechanism in terms of the competition of theoretical orientations in mind, one may state that science approaches the absolute truth due to the competition of different theories accepted by different groups of people. Their rivalry is the best means to achieve scientific progress, i.e., to achieve a theory which is more essentially true than its predecessors. The goal of the ideal, typical investigator is to explain phenomena. The goals of actual investigators may deviate from this: they aim at persuading people, at gaining social prestige, etc. But what is really important is that there is in science a mechanism which transforms particular actions governed by such objectives into attaining more and more adequate essential truth. Essential truth is not an 'aim of science'; it is an objective effect of making science. It is, however, its 'internal' effect. External effects of scientific practice will be discussed in the next chapter.

CHAPTER 14

THE SOCIAL CONTEXT OF SCIENCE

1. THE INTERNAL STRUCTURE OF A COGNITIVE PROCESS

Let us try to summarize the cognitive process taking place within a scientific community. According to the traditional Marxist formulae, the cognitive process occurs in two stages: (1) 'from empirical reality to the abstract', (2) 'from the abstract to (understood) empirical reality. The first stage consists in idealization, while the other in concretization: this is the main idea of methodological parts of this book. Now let us consider what has been added to this idea in the last chapter due to accounting for the cognitive role of a scientific comunity.

(I) A typical investigator assumes a given categorization of the world and a given ontological perspective which is composed of principles of essential stratification and principles of ontic connections. The principles of essential stratification allow the class of images of the essential structure, while the principles of ontological connections allow the class of dependencies which may be the case. For instance, mechanicism contained the following principles: (a) for any physical magnitude, its essential factors characterize the states of molecules composing the bodies defined by this magnitude, (b) for any physical magnitude, all dependencies which hold between any magnitude and its essential factors consist in either attraction or repulsion between the bodies characterized by these magnitudes.

(II) An investigator selects one image of the essential structure of the magnitude he is interested in among the class of images allowed by his ontological perspective. Reasons for this choice may be empirical of theoretical in nature (see Chapter 8 and Chapter 9).

(III) Once the image of the essential structure for a given magnitude has been established, an investigator starts determining the dependence which holds between the magnitude distinguished and factors considered to be principal for it. The ontological perspective defines the type of this dependence, whereas the concrete form of it can be discovered by empirical research or purely theoretical considerations. The idealizational statement stating the dependence in question is next concretized.

(IV) The theory obtained in this way is tested. If the result is negative, the investigator concludes that the dependence provided by the final concretization has been established incorrectly. A new form of the correctional or directional function is proposed. If no agreement of the new final concretization with empirical data is obtained, the attempts at correction go on further: concretizations of higher and higher order are being replaced by new ones. Finally, the dependence proposed in the initial idealizational statement is revised and new process of concretization follows until the final concretization is obtained. The new form of the theory is tested again. If it agrees with actual data, then the modified project is accepted, if not, further-reaching revision is necessary.

(V) The investigators who control the project (which may happen immediately after it is proposed, or years later, when it functions in science) come to the conclusion that the incompatibility with experience results from deeper sources: from an improper image of an essential structure of a definite magnitude. First they try to change the repertory of secondary factors. For instance, they discover new minor factors, so they abstract former theorems of the theory even further (they attach new idealizing assumptions) and remove these assumptions by way of concretization. Hence, a new version of the theory corresponds dialectically to the older one – a regularity is upheld (with an increased number of idealizing assumptions) and it is modified with respect to an increased number of secondary factors (Nowakowa (1)). A new ultimate concretization obtained in this way is then tested. In case the results are positive, a new theory which dialectically corresponds to the old one is accepted in a scientific community (although it had previously been rejected). In case the results are negative, the revision goes further – the repertory of the main factors changes (the perception of 'essence' changes compared to the primary project). A new image of essential structure is established, basically different from the previous one (i.e. with a different set of main factors), but some former significant factors are still viewed as such. A new law is established and the new concretizations obtained; a new theory is built and it corresponds in a weaker sense to the old one (Nowak (11)) – some correctional functions which appear in the concretizations of the new theory are identical to some dependencies which occur in the former theory's dependencies. And this theory, which corresponds in this sense, i.e. 'derivatively' to the older one, is tested again. If it fails, the way towards further modifications is blocked in principle – the project turns out to be a 'dead end' street since even far reaching modifications (the ones which correspond derivatively) fail. A scientific group waits for new projects then, and they become material for modifications according to the same scheme: first one changes the form of dependencies in concretizations, then in the law, then one revises the repertory of the secondary factors, obtaining a theory which corresponds basically to the older one, finally, the repertory of the main factors is changed and a derivatively corresponding theory is obtained.

(VI) If a number of projects based on the same ontological perspective turn out to be 'barren' in the above sense, i.e. none of successive modifications, which perhaps will take years to be done, does not result in compatibility with experiments, a belief grows in a scientific group that former theories proved to be incompatible with experience not by accident (in spite of their makers conviction that the compatibility does occur), and that it was not a coincidence that all modifications have been rejected after some time. Responsibility for that is slowly heaped on the mistaken ontological perspective assumed by all the projects which result in successive corrections of the growing size (from the change of correctional functions to the revision of the principal factors). A conviction becomes widespread that the accepted ontological perspective (which, by the way, science becomes aware of no sooner than now) is wrong and calls for replacement. This is a period when outstanding specialists start engaging in philosophical discussions. What happens is either that other ontological perspectives drawn by philosophy are accepted in science, or new perspectives are created within science itself, that is, an outstanding specialist proposes a new theory which assumes an ontological perspective different from the previous one.

It is possible that after revisions of ontological perspectives, the revision of ontological categorization of the world also takes place. But this is rather connected with a change of the language of science because it is a language which presupposes a definite categorization of the world.

Hence, the structure of scientific activity of a typical investigator is a synchronic schema which undergoes changes due to the cooperation and efforts of other investigators (representing different theoretical orientations). The social nature of making science sets scientific structures in motion.

The internal structure of a cognitive process may be then presented as in Figure 12.

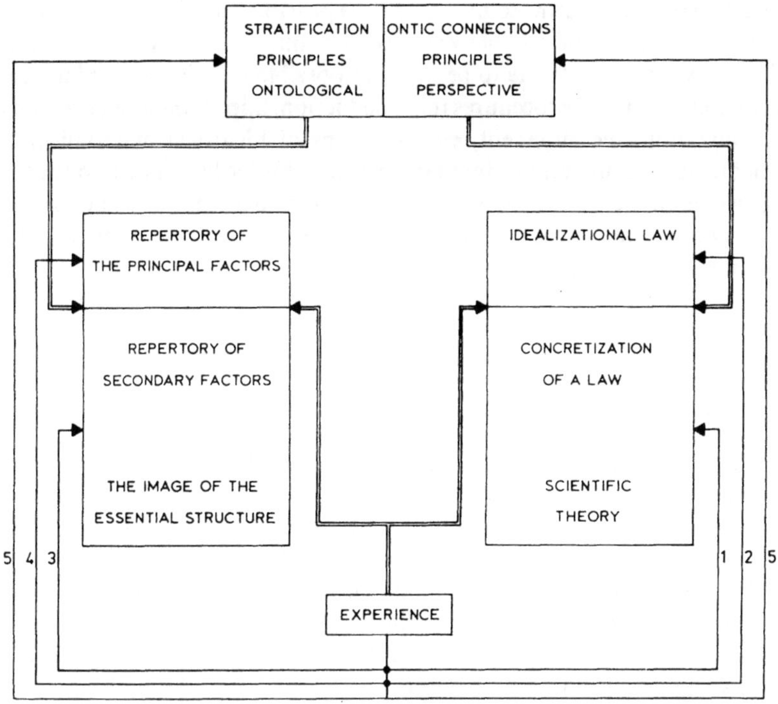

Fig. 12. The symbol ⇒ stands for the relation of generation (of a *class* of cognitive constructs), while → stands for the relation of controlling or testing. The number at an arrow points out the order of procedures improving the given construct (concretization, a law, an image of the essential structure, etc.) after the negative results of testing.

2. THE PROBLEM OF THE SOCIAL CONTEXT OF THE INTERNAL STRUCTURE OF A COGNITIVE PROCESS

Having established the image of the internal structure of a cognitive process, one may raise the question in what way this structure is connected with its 'social environment'. There are numerous works in Marxist literature where this connection is claimed to be the most important for understanding science but those ideas are usually obscure and declarative. The real problem is, in my opinion, to understand in what way the social influence upon the internal structure of scientific cognition is possible at all. Certainly, some Marxists assure us that it is impossible to reconstruct the internal structure of scientific process without the means of historical materialism. But it seems to me that this *is* possible and that doing this one may reach a piece of reliable knowledge about what is going on within science. What seems to me to be really impossible is giving an explanation of concrete scientific procedures (e.g., deduction, approximation, statistical concretization, measurement, etc.) in terms of historical materialism. If someone is of a different opinion concerning this let him try to do that.

It seems to me that there is a possibility to understand in what way the social context influences the structure of science. In order to show this it is necessary to introduce some axiological concepts.

3. SOME AXIOLOGICAL NOTIONS

(a) Value and Factor

The starting point of our considerations (see also Nowak (12)) is the idea of unity of being and duty (value). It is interpreted here as follows: the same schemes are common to basic ontological and axiological notions. Let us try to show this with the aid of some examples (see also Nowak (13)).

The initial concept is that of preference. We shall be concerned, however, not with the preferences of an individual, but with the preferences of a 'social subject' who expresses a standard of valuation that is fixed in a given social group (a class, section of a class, or – in the limiting case – society as a whole). The preference ascribed to that social subject is not any average of the preferences which the various individuals have, but a preference which accounts for the fact that the social agent's actions are necessary for his group holding in the social structure the position it actually holds. In this interpretation, the capitalist as described in *Capital* is such a social subject. He has only one preference, namely the maximization of profit. Marx ascribed it to him not because most capitalists are guided by such a preference (it being well known that no real capitalist is guided in his actions by a single preference only), but because the maximization of profit by the individual capitalist is necessary for that class to be dominant.

We shall not, however, concentrate on these substantial conditions imposed upon the concept of preference since they are not essential for the problems with which the present chapter is concerned. Moreover, certain formal conditions, also are imposed upon the concept of preference. Preference, is a relation defined on the set W of states of affairs it is also an antisymmetric and transitive relation (von Wright (1), p. 12). Hence, if a state of things p is preferred to a state of things q then it is not true that q is preferred to p; and if a state of things is preferred to a second one, and a second one to a third one, then the first one is preferred to the third one.

The concept of preference is used in the following definition of the relation of being equivaluable: p and q are *equivaluable* if and only if it is not true that p is preferred to q and it is not true that q is preferred to p. This relation is reflexive (every state of things is equivaluable to itself), symmetric (if a state of things is equivaluable to another, then the latter is equivaluable to the former), and transitive. It is thus an equivalence relation which divides the set W of states of things into the classes Z, Z', Z'', . . . of pairwise equivaluable states of things. Each of these classes contains equivaluable states of things (i.e. equivaluable for a given person): further, no two states of things contained in different classes are equivaluable, but one is always preferred to the other. Now that set of the classes of equivaluable states of things is a *value* for a given social subject. These classes of equivaluable states of things are *cases of value*.

All values are, as it seems, opposition-based families of sets of states of things, i.e., such which contain both a maximum and a minimum under the relation that orders a given family. In other words, a value is a set $V = \{Z_{max}, Z, ..., Z', Z_{min}\}$ such that Z_{max} and Z_{min} are, respectively, the set of those states of which are maximally preferred and the set of those which are minimally preferred (by a given social subject). We assume that between the *maximum case* Z_{max} and the *minimum case* Z_{min} of a given value there is an *indifferent case* Z_{ind} which contains those states of affairs which are *axiologically neutral*[1]. Every positive case of a given value, i.e. every set of the states of affairs in a given value which precedes (under the relation of general preference) its indifferent case will be termed a *good*. Every negative case of a given value, i.e., a set of the states of affairs which are preceded (under that relation) by the indifferent case of that value will be termed an *evil*. The structure of a value V may be represented thus: $V = \{Z_{max}, Z, ..., Z', Z_{ind}, Z'', ..., Z_{min}\}$ where $Z_{max}, ..., Z'$ are goods, and $Z'', ..., Z_{min}$ are evils.

Compare now the above construction of the concept of value with the usual construction of the concept of magnitude (factor) (e.g., Ajdukiewicz (1)).

Given a set of U objects and a precedence relation R defined on that set and assumed to be antisymmetric and transitive, we define the indistinguishability relation thus: two objects are indistinguishable if and only if it is not true that one of them precedes the former under R. If we assume that the relation of indistinguishability is transitive, then we can prove that it is an equivalence relation, which thus divides the set U into classes of objects which are pairwise indistinguishable under R. A family of such classes is a *magnitude (factor)*. In a special case this family may be ordered as, according to the above conjecture, is the case with values. When we compare this construction of the concept of magnitude (factor) with the construction of the concept of value we see a complete formal analogy between the two: the process of construction is precisely the same at each step. In other words, we might say that the ontological structure of value and that of magnitude (factor) are the same.

(b) Axiological Structure and Essential Structure

The social subject under consideration is marked by a number of preferences rather than by a single one. This is why a number of values may be ascribed to him. Now, when we evaluate a state of affairs, then in the set of all values we single out those which are important (significant, relevant) for that state of things, and then a hierarchy is established even among these important values. Some values are singled out as the fundamental ones which must unconditionally be taken into consideration when a given case is being evaluated, whereas others are treated as secondary, which can merely modify or complete the evaluation based on fundamental values. Thus, for instance, when we evaluate physicians we take the effectiveness of their treatment to be the fundamental value. That evaluation may also be modified from the point of view of such values as behaviour towards the patient so as to become a joint evaluation on which the decision which physician to choose can be based.

In other words, in a moral or aesthetic discourse some criteria of domination of values are assumed. What these criteria are is an empirical issue which will be disregarded here. The important point is that there *are* some criteria which enable us to accept that in the evaluation of physicians the value of the effectiveness of treatment dominates the value of good bedside manner and that in the evaluation of students the value of intelligence dominates the value of sociability.

Let us accordingly assume that the criteria of the domination of values have been fixed for a given social subject (while they may be different for other social subjects). Let us also assume that in evaluating the states of affairs of type K the social subject under consideration holds the values V, S_k, ..., S_1 to be important, and that these values are such that V dominates S_k, S dominates S_{k-1}, etc. The *axiological structure* of the states of affairs of type K for the social subject under consideration becomes the following hierarchy of values:

(k) V,
$(k-1)$ V, S_k,
... ...
(1) $V, S_k, \ldots, S_2$,
(0) $V, S_k, \ldots, S_2, S_1$,

which includes zero, first, etc., *levels of domination*. The zero level will be called the *surface level*, because it includes all those values which the subject holds to be important for the cases of type K. The kth level will be called the *internal level*, because it includes only the *fundamental value* V. Evaluating cases of the type K the agent disregards all the *secondary values* $S_k, \ldots, S_1$ and thus assumes counterfactually that the state of affairs being evaluated is axiologically neutral from the point of view of those secondary values, i.e. that it forms an indifferent case of those values. Obviously, there may be more fundamental values (equivalent to one another, that is, such that one does not dominate the other), likewise, at a given level there may be more pairwise equivalent secondary values.

S. Morawski has adopted a 'system of criteria for the evaluation' of works of art (Morawski (1), pp. 330 ff.) by which we can interpret a special case of the axiological structure in the sense defined above. That system includes artistic criteria: the constitutive value of a work of art, its novelty, and its originality, and extra-artistic ones, such as moral, political and the like. The former are dominant. "There is no reason to reject an extra-artistic evaluation of works of art. The art theorist may only demand that the division into these two different types of criteria of evaluation be respected, and that a given critic should always realize what criteria he uses. There is also no doubt that the art theorist will argue, not without justification, for the application to works of art of those criteria which are proper to it, that is, artistic ones." Artistic values are here explicitly treated as dominant over extra-artistic ones, and priority is given to an evaluation based on artistic values, extra-artistic ones being disregarded. Among the latter Morawski gives the pride of place to "such an ideological qualification which results from the mimetic tissue of the work"; he emphasises that "those ideological elements which are mainly connected to the mimetic transformation of reality in art may play considerable role in the scaling of works of art". The other extra-artistic criterion which he seems to place at the same level is "the evaluation of a work of art from the point of view of disalienation." He gives a lower status to the other two criteria: an ideological qualification "of works of art which desregards the artistic structure of a given work, its inner elements, and their mutual relationships", and the criterion of comprehensibility of art.

Thus, the fundamental values, in the axiological structure which Morawski adopts for the works of art, are: the constitutive values of a given work (*k*), its novelty (*n*), and its originality (*o*). The extra-artistic values are treated as secondary: among them we find those on the higher level: the ideological values of the artistic structure of a given work (*m*), and disalienation (*d*), and those on the lower level: the ideological expression which disregards the artistic structure of a given work (*m'*) and comprehensibility (*z*). Hence the axiological structure of a work of art, as adopted by Morawski, is this:

$$k, n, o,$$
$$k, n, o, m, d,$$
$$k, n, o, m, d, m', z.$$

Here, again, we can see a strict analogy between this axiological structure and the fundamental concepts of ontology. As I have striven to demonstrate, in Marxian ontology the basic concept is that of the essential structure of a given factor (magnitude). Assume namely that the significant magnitudes for a magnitude *F* (i.e., these magnitudes which affect *F*) are $H, p_k, \ldots, p_1$. But the magnitudes which are elements of that set are marked by a certain hierarchy: some of them are more significant for *F* than the others are (i.e. they affect *F* more than the others do). Assume that *H* is more significant for *F* than P_k is, that p_k is more significant for *F* than p_{k-1} is, etc. The hierarchy of magnitudes

(k)	H
$(k-1)$	H, p_k
...	...
(0)	$H, p_k, \ldots, p_1$

is the essential structure of the magnitude *F*, consisting of zero (surface), first, ..., *k*th (inner) levels of significance, where *H* is the principal factor and $p_k, \ldots, p_1$ are secondary factors.

As can be seen, the concept of axiological structure is formally fully analogical to that of essential structure. As the reconstruction of the essential structure is a tacit assumption which underlies all cognition, as the reconstruction of axiological structure is a tacit assumption – in the same way – which underlies all valuation.

4. EVALUATION AND SCIENCE

The above fact – namely that value is a magnitude as well (to be more precise – magnitude may form a "component" of value) – has further consequences. If the relation of value domination (for these values which are magnitudes as well) coincides with the relation of greater significance, then axiological structure may be an essential structure as well. Main factors (magnitudes) may be principal values as well. The very preferences of a scientist may therefore lead him towards the essential structure for explained phenomena. And this is the most important reason for justifying valuation in science, or, more modestly, at least in the humanities. No simple relations hold, to be sure. It is not true that valuations always lead one towards establishing essential structure. It is also not true that axiological structure never coincides with respective essential structure. It is true, however, that the axiological structure a scientist accepts (the structure itself is suggested by the preferences of a social group of scientists) may coincide at times with the proper significance structure of investigated phenomena.

Let us recall a capitalist in Marx's view. He assumed that the only aim a capitalist has is the maximalization of profit. This idealizing assumption constitutes an ideal type of a capitalist (industrial one), as well as a negative moral ideal from Marx's point of view. The fact, that real capitalists approach the ideal type was seen by Marx as a demonstration of 'wolf-like nature' of the capitalists. J. Robinson says that a capitalist in the Marxian view is not a "bold and restless industrialist" (as was the case in Marshall's theory) but a "greedy and cunning niggard" (Robinson (1)). It is clear that the very preferences of the author of *Capital* suggested a construction of an ideal type of the capitalist to him, and that this is at the same time a negative ideal on the grounds of his morality. If we agree that Marxian theory is the true theory of the capitalist mode of production for free competition period, then we have to admit that Marx's evaluations suggested the idea which was cognitively valid. This is an example of cognitively valid – as far as we know – ideal type, whose construction has been suggested by valuations a scientist accepted. Let us consider an example of a significance structure which is an axiological structure at the same time.

Marx presented in the first volume of *Capital* capitalist exploitation in its pure form. He abstracted from everything which would interfere with the presentation of expropriation of surplus product by the capitalist: from the fact that industrial capitalist allows for a commercial profit of a commercial capitalist, percentage for a financial capitalist, rent for a landowner, etc. He has, therefore, constructed the following significance structure:

$$W,$$
$$W, Z_h,$$
$$W, Z_h, P,$$
$$W, Z_h, P, R,$$

where W is a surplus product obtained by an industrial capitalist from the worker, Z_h is a commercial profit, P denotes percentage, and R is the land rent. The main magnitude (factor) is a component of principal value: capitalist exploitation. It consists of a social situation in which surplus value is being appropriated. Needless to say, capitalist exploitation was the principal evil of the system. Other forms of appropriation were, so to say, derivative evils. Presenting the social situation which consists of appropriating surplus product as $S(W)$ and using analogical terminology for the other cases we arrive at the following axiological structure:

$$S(W),$$
$$S(W), S(Z_h),$$
$$S[W], S(Z_h), S(P),$$
$$S(W), S(Z_h), S(P), S(R).$$

It is clear that its 'component' is the essential structure presented above, the only difference may consist of the minor factors being equally dominant for Marx, hence the above axiological structure would be equivalent for him to the one where appropriation of percentage ($S(P)$) was on the second level and not appropriation of commercial profit. Then the choice of the first of the above alternatives as the basis for constructing significance structure would be prompted by cognitive reasons alone. The most important fact is that surplus product is the main factor (magnitude) – more significant than the other magnitudes in question, and the situation of appropriating surplus product is a (negative) principal value – dominating all the remaining values. Hence trying to demonstrate capitalist exploitation Marx revealed – at least this is what every Marxist thinks – the essence of capitalist system. And this is why he demonstrated how much he despised capitalism. Valuations led him towards a scientific discovery of the basic range, to the greatest of possible discoveries: to the revaluation of the essence of an investigated phenomenon.

Needless to say valuations which lead a scientist towards an essential structure do not play any other role apart from inspiration. First of all they do not justify the accepted theses of a cognitive kind. Although valuation may lead towards the reconstruction of the essential structure of investigated phenomena, it does not provide a justification of the thesis that it is so in a given case.

5. CLASSES, VALUES, COGNITION

One should not, of course, stop at a statement that essential structures do coincide with axiological ones at times. The answer to the question as to when it is so belongs to sociology of knowledge. And this is so since – as has been pointed out before – all axiological concepts are relative to a social group, to be more precise – to an ideal social subject representing the group. Therefore the question concerning when axiological structures coincide with those which are essential may be presented in its fuller form as follows: when do axiological structures determined by social preferences of the group, with which a scientist identifies himself, coincide with the essential structure for a particular phenomena? The problem is to determine the relation between the position of groups in society and axiological structures provided by social preferences in their frameworks, and then to determine the relation between the latter and essential structures for investigated phenomena. The crucial role between social groups shall, of course, be played by social class. Idealizational reconstruction of the relation which is assumed in historical materialism with respect to this, should, therefore, start with the assumption that there exist classes alone: other social groups should be introduced gradually while complicating the assumed image of social structure.

This reconstruction is beyond the scope of both my plans and abilities. It is a separate, important, quite virgin subject. I think one of the reasons the Marxists are slow with respect to this subject is the improper posing of the principal problem of the Marxist sociology of knowledge. It is being considered as the question: what are the relations between the social position of a given groups and cognitive results obtained by its representatives. Many profound answers may be provided to this problem, but they are vague. If one wants to answer more precisely, profoundity disappears, and the answers turn out to be either trivial or clearly false. How can one think of establishing regularities relating the position of a class or its mind (for instance values it professes) to the contents of a theory built by the people considered to be the representatives of this class? It is clear that no clearcut relations exist.

The basic problem of sociology of knowledge may be formulated in other words: what relation exists between axiological structures accepted in a given group and essential structures assumed by a scientist representing this group. Or more generally: between systems of values accepted by the group and the systems of magnitudes assumed by a scientist representing this group. The theorems which might be established by sociology of knowledge, would therefore link the position of the group not to definite theories but the position of the group along with the systems of values functioning within it (or some definite axiological structures) to the families (or types) of theories. It would not presuppose, therefore, that if a scientist represents such a group, this theory will be accepted, but a class of acceptable theories would be outlined (the class of theories which are based on an assumption about some definite essential differentiation of reality).

The reconstruction of these theorems of historical materialism which belong to sociology of knowledge is therefore a separate task. I would like to limit myself to the outline of reconstruction of just one theorem – perhaps the most often discussed. The theorem says that the progressive class has more cognitive possibilities than the conservative. The starting point for the Marxist sociology of knowledge suggested above should allow us to understand why this theorem is true. The progressive class searches for the factors of social reality which lead to changes. These are main factors, and at the same time principal values from its point of view. The theories built by the representatives of the progressive class may be different from each other, but they have one general common property – they do not abstract from 'change-introducing' factors considered to be main ones, and remaining the principal values for this class. The theories present therefore the mechanisms of changes, they are dynamic ones. The conservative classes select the factors which maintain *status quo* as the main ones. These are the factors which the scientific representatives of these classes consider main ones. The theories they built may differ from each other as far as the contents of theorems is concerned, but they have one common feature – they do not abstract from the stabilizing factors considered to be main ones, and remaining principal values for these classes. Now, if a given class is *de facto* progressive in the meaning historical materialism ascribes to it (it expresses 'the movement of history' as is often said) then idealizational theorems abstracting from stabilizing factors and not from 'change-introducing' ones are closer to reality than idealizational theorems abstracting from 'change-introducing' ones, not from stabilizing ones. To be more precise: the approximations of the former may be fulfilled, while the approximations of the latter are definitely not. What is more – the future will show the truth of the former and falsity of the latter.

NOTE

[1] I owe this remark to J. Kmita

CHAPTER 15

THE SOCIAL REASON FOR MAKING SCIENCE

1. CLOSED AND OPEN EPISTEMOLOGIES

As it has been shown (see Chapter 13), there are different competing theoretical orientations within a scientific community. Each of them develops itself, due to the competition with the others, according to the principle of dialectical correspondence. This is why the basic question is what criterion points out which of the rival theoretical orientations starts with the image of the essential structure that is basically similar to the structure itself. If this orientation is called a proper one, then the question can be formulated as follows: which of the competing theoretical orientations is the proper one?

Different epistemological doctrines can be distinguished depending on how they are able to solve the problem of the criterion of a proper orientation. Closed epistemologies are inclined towards the search for this criterion within the framework of scientific cognition itself – e.g. in coincidence with experiments, internal coherence of a theory, etc. Open epistemologies suggest non-cognitive criteria. For Marxism, I believe, this is the criterion of practice.

2. THE MEANING OF THE FORMULA: "*PRAXIS IS THE CRITERION OF TRUTH*".

(a) Difficulties of the Current Interpretation

A current interpretation of the formula "practice is the criterion of truthfulness of human knowledge" suggests that coincidence and compatibility with practice is the basic criterion of truthfulness of statements about the world. This interpretation is open to grave charges, I think. First comes the historical argument. It is hard to understand why Engels, on the one hand, thought that a criterion of practice is a peculiarity of the Marxist doctrine, while, on the other hand, failed to mention this criterion when writing about the growth of scientific theories, although he often mentioned compatibility with experiments (Engels (3), pp. 256–257). The second argument concerns not historical facts but substantial adequacy of this interpretation. It is hard to deny that in the empirical sciences, like physics, astronomy, etc., the criterion of truthfulness of particular statements is their compatibility with experiments – the highest precision of explanation of the largest possible amount of discovered facts. It does not seen to me that practical efficiency of these statements plays an important role for the physicists: their truthfulness is already decided when technicians use them for practical designs. Moreover, technicians use them exactly because they trust truthfulness demonstrated experimentally within theoretical sciences.

(d) On Some Misinterpretations of the Criterion of Practice

As the current interpretation suffers hard criticism, it is also very often that we encounter a tendency to identify the criterion of practice with compatibility with experiments. It is said that the Marxist criterion of practice assumes prediction of observable, phenomena, and if the latter occur it proves that knowledge in question is true, but when they fail to appear – it is false. This intellectual tendency overcomes the second charge in a purely verbal manner. The first charge is still untouched, though – why did the founders of Marxism differentiate between the criterion of practice and compatibility with experience?

(c) Practice as a Criterion of Relative Truthfulness

Both the current interpretation of the above formula and the tendency to identify the criterion of practice with a criterion of compatibility with experience, remain within the same intellectual tendency. It is founded on a belief that the formula "practice is the criterion of truth" includes truth in the classical meaning of the term. However, I think that this formula means a relative truthfulness. And if the interpretation of the concept of relative truth we assume is right, then it follows that social practice is the criterion of distinguishing the proper theoretical orientation. Criterion does not serve to decide whether particular sentences report facts or not, but to decide which of the rival theoretical orientations assumes the image of essential structure which is basically similar to the structure of a magnitude under investigation itself: finally, in which of these orientations one can expect to approach the absolute truth.

Let us now examine more closely the working of the criterion of practice. It will require an explanation of the way in which scientific knowledge can be applied in practice.

(d) Application of Scientific Knowledge

The first stage of the application of scientific knowledge consists in deduction optimalizational statements:

(1) if the factors $q_1, \ldots, q_m$ do not act, and if the conditions U_i occur, then performance of the action A_j results in the optimum state of affairs r_{ij}.

Optimum is a state of affairs which consists in the fact that a given magnitude assumes a value for some object while this state of affairs is an ideal according to presupposed evaluations (see Chapter 14, Section 3(a)).

The second stage consists in applying optimalizational statements in order to make decisions. Let us consider the possible conditions $U_1, \ldots, U_p$ and actions $A_1, \ldots, A_n$ of the results $r_{11}, \ldots, r_{np}$:

$$\begin{array}{c|l} & U_1 \;\ldots\; U_k \;\ldots\; U_p \\ \hline A_1 & r_{11} \;\ldots\; r_{1k} \;\ldots\; r_{1p} \\ \vdots & \ldots\ldots\ldots\ldots\ldots\ldots \\ A_l & r_{l1} \;\ldots\; r_{lk} \;\ldots\; r_{lp} \\ \vdots & \ldots\ldots\ldots\ldots\ldots\ldots\ldots\ldots \\ A_n & r_{n1} \;\ldots\; r_{nk} \;\ldots\; r_{np} \end{array}$$

Let the states at affairs $r_{11}, \ldots, r_{np}$ consist in the fact that a magnitude assumes definite values for objects in question. In each column k there is the result r_{lk} which precedes all the other results being an optimum state of affairs for the conditions U_k. The role of optimalizational statements is to show how to actualize it as the most desirable state of affairs.

(e) Relative Truth and Practice

Let us have the idealizational statement of the standard form T^m and its consequence (1). The following statement:

(2) if the factors $q_1, \ldots, q_m$ act weakly and the conditions U_i occur, then performance of the action A_j results in the state of affairs r'_{ij} close to r_{ij},

will be the approximation of (1). If we base our decision on optimalizational statement (2) and the actual result of our action is really close to optimum r_{ij}, then we are ready to recognize that the statement (1) furnishes us with a basically proper means to achieve our goal r_{ij}; deviations (since

r'_{ij} and not exactly r_{ij} has been achieved) are explained by the influence of secondary factors $q_1, \ldots, q_m$ which have not been eliminated entirely. The statement (1) is, in turn, a consequence of the idealizational law T^m. Practical efficiency of the statement (1) and of other optimalizational statements entailed by T^m testifies indirectly to the fact that a factor, accounted for in this law, is more essential than factors which had been disregarded, that is, it is the principal factor. Thus the practical efficiency of optimalizational consequences of T^m testifies to its representing regularity and hence being a relative truth. Were it not the case, no consequences which point to basically proper means of actualizing the aims would follow from T^m (basically – that is, accounting for the deviations caused by uneliminated influence of secondary factors). Therefore, efficiency of a practice based on a definite theory means that the theory had recognized properly the essence of investigated reality.

Thus within this theoretical orientation which includes such an effective theory one may expect the process of approaching the absolute truth.

3. AMELIORATION OF SOCIAL PRACTICE

(a) The Problem of the Marxist Theory of Science

At the beginning of this part of the book (see Chapter 7, Sections 3(d, e)). I have presented the problem of the whole theory of science based on the Marxist tradition in the following way: how to explain that as a result of actions motivated by a desire to discover the internal structure of facts (with the aid of the method of idealization) the amelioration of social practice follows? This is the problem which unifies the idealizational interpretation of the Marxist methodology (theory of cognitive activities) and the theory of cognitive practice. So far we have presented a kind of systematic methodology (Chapters 8–12) and then we have demonstrated how competition of different theoretical orientations results in progress of science (Chapter 13) which consists in a transition from less to more adequate relative truths. At present I shall try to demonstrate that there is a definite dependence between obtaining increasingly adequate relative truths and practical efficiency of their optimalizational consequences, hence – between the approaching absolute truth and efficiency of social practice based on it.

(b) Relative Truthfulness and Practical Adequacy

Let us compare two relative truths:

(T^m) if $U(x)$ and $q_1(x) = 0$ and ... and $q_{m-1}(x) = 0$ and $q_m(x) = 0$, then $F(x) = k_m[B_1(x), ..., B_n(x)]$

(T^{m-1}) if $U(x)$ and $q_1(x) = 0$ and ... and $q_{m-1}(x) = 0$ and $q_m(x) \neq 0$, then $F(x) = k_{m-1}\,[B_1(x), ..., B_n(x), q_m(x)]$.

The latter is more essentially true (or more adequate), for it represents the form of manifestation of the regularity, while the former is only the regularity itself. Therefore, T^{m-1} is closer to the absolute truth representing the surface form of manifestation of the regularity. This is why the image of the essential structure of the magnitude F assumed by T^m includes at least the principal factor for F, while the image of this structure assumed by T^{m-1} includes not only the principal factors but also actual secondary factor q_m.

Let us assume that the statement T^{m-1} is a more essentially true statement than T^m. Let us also assume that T^m has the optimalizational consequence (1), while the optimalizational consequence of the relative truth T^{m-1} is the statement:

(3) if the factors $q_1, ..., q_{m-1}$ do not work and q_m does, and if conditions U_i occur, then performance of the action A_j and the action a_m results in occurrence of the optimum state of affairs r_{ij}.

The action a_m may consist in doing something which compensates for the influence of the secondary factor q_m. On the basis of approximation of the statement (3) one can actualize a state of affairs r''_{ij} which is closer to intended r_{ij} than r'_{ij} allowed by the approximation of (1). It follows from the fact that (3) admits additional act a_m whose result makes up for the disturbances caused by the secondary factor q_m. It can be termed that (3) is practically more efficient than (1), because practice based on the former is more efficient than practice based on the latter.

It is possible, I think, to express it in a more general form: it t' is a more essentially true relative truth than t, and z' is an optimalizational consequence of t', while z is an optimalizational consequence of t, then z' is practically more efficient than z. And since the growth of knowledge consists, as we have seen in Chapter 13, in passing from one relative truth to another, more adequate (more essentially true) one, growth of science results in increasingly efficient optimalizational statements. Finally, social (non-cognitive) praxis based on the latter becomes increasingly efficient. To sum it up: the progress of science ameliorates social praxis.

(c) The Aim of an Investigator and the Global Effect of the Cognitive Process

Hence the proposed solution of the problem of the Marxist theory of science can be formulated thus: the subjective aim of a typical investigator is to explain facts (with the aid of the method of idealization); subjective aims of various investigators who represent different theoretical orientations jointly lead towards the absolute truth; further global effect of cognitive action undertaken by investigators is that social (non-cognitive) practice based on gradually corrected scientific knowledge becomes increasingly effective. Therefore, striving towards explanation of the facts ultimately results in amelioration of non-cognitive practice. This is why science as a social institution is maintained by society.

I would like to finally add that this answer to the basic problem of the Marxist theory of science is based on an interpretation of this theory itself. There are certainly possible interpretations of another kinds emphasizing different threads in the Marxist tradition. In particular: a different partition of the Marxist theory of science, a different relationship between them, another type of basic procedures of scientific conduct, etc. Nobody is able to show in advance "the proper interpretation" before careful analyses of the traditional Marxist ideas and the scientific practice of the founders of Marxism are done. What is necessary is the plurality of these interpretations – nobody knows a priori which of them will turn out to be the best way of explaining the structure of science.

PART IV

APPLICATION: AN ANALYSIS OF SOME CRITICISMS OF MARX

Part I of this book has proved, I think, that Carl Marx had applied a system of definite methodological principles in his scientific work that seem to be rather original, as it was stated in Part II, in comparison with methodological rules offered by modern trends in the philosophy of science. In Part III, I have tried to develop the Marxian principles in order to demonstrate the possibility of building a theory of science based on these principles. At present, I would like to come back to Marx's writings and to show that their methodological novelty has not been understood at all, as is manifested, for instance, in numerous criticisms of the Marxian concepts. Many critics, both left and right, do not notice the original methodological assumptions underlying Marx's social theories and, what is more, their criticisms seem to fail just because of this. Let us consider some of them from the point of view of the Marxian methodology interpreted in the idealizational manner developed in this book.

CHAPTER 16

THE LAST RESORT

1. HISTORICAL MATERIALISM: A SINGLE-FACTOR THEORY?

It is a rule in sociological studies, especially the non-Marxist ones, to classify historical materialism as the so-called single-factor theory. These are the theories which try to explain all social phenomena by means of one kind of factors alone, e.g., geographical, demographical, etc.,ones. Historical materialism is supposed to elevate the economic factor to this position (it is by no means clear what is the exact meaning of the 'economic factor' in the studies of these critics of Marxism. I shall not go further into it, since my only purpose is to show how baseless this kind of criticism is from the methodological point of view). The following statements are therefore considered sufficient to oppose the Marxist theory of social development:

> The relations between the economic and non-economic phenomena are complex, and various social phenomena are related to the economic ones in varying degrees. It has been established . . . that such relations are by no means complete. There is no such non-economic phenomenon whose nature or changes might be explained by economic factor alone. Economic phenomena should not be treated as something that conditions the other phenomena, since they themselves may be shaped by the other ones. The economic factor cannot therefore by treated as a 'cause' and the other phenomena as 'effects'. We are fully entitled to the view that the other phenomena are independent magnitudes and the economic phenomena are their functions (Sorokin (1), pp. 115-116).

As may be easily observed two basic theses are hereby ascribed to historical materialism: (1) for every social phenomenon there may be found such an 'economic factor' which is either direct or indirect sufficient condition for this phenomenon to occur, (2) economic phenomena are not influenced by non-economic ones. Therefore a well-known Engelsian saying that "economic necessity always – in the last resort – finds its way" is being intepreted as if economic factors shaped all the other ones in a causal way and did not require explanation themselves within the causative chain. If historical materialism, however, was understood in such a way, then it should be considered an internally contradictory theory. Engels has written:

We think that economic conditions determine in the last resort historical development . . . . Political, legal, philosophical, religious, literary and artistic development is based on economic development. But all these factors mutually interact and they also influence the economic base. It is not the case therefore, that the economic situation alone is the cause, the only active element, while everything else is just a passive effect. Quite the contrary, there is always a mutual interaction on the basis of economic necessity which in the last resort finds its way . . . . It is not the case, then, as many imagine to their utmost comfort, that economic situation works automatically, to the contrary, people make their own history, but all this happens in a given milieu which conditions their actions, on the bases of pre-existing factual relations, out of which the economic ones (no matter how much they may be influenced by the other – political, ideological – relations) decide in the last resort and make a red thread visible everywhere, the only guide in understanding historical process (Engels (1), p. 482).

It is evident that the above quotation includes the sentences contradicting the popular interpretation of historical materialism (exemplified by Sorokin). Engels says that not only economic phenomena play an 'active role' in social life and that the 'economic conditions' are themselves influenced back by the other factors. If so, the above interpretation of the notion of the last resort is inadequate. The thesis that 'economic necessity in the last resort always finds its way' does not mean that 'economic factor' is the final sufficient condition for all social phenomena. How are we, then, to understand this thesis?

2. THE NOTION OF THE LAST RESORT

The basic thesis of the Marxian model of science says that the basic determinants of the investigated phenomena are accounted for in the initial idealizational laws. These laws eliminate secondary factors and, therefore, their possible influence upon principal factors. For it may happen that secondary factors not only influence the investigated magnitudes but also principal factors for these magnitudes. Sometimes it happens that factors belonging to the essential structure of the given magnitude influence each other (see Brzeziński *et al.* (1)); in cases of the type in reality we deal with the complex system of mutual dependences. But all this is eliminated on the level of principal factors – secondary factors and their back influence on the principal ones are responsible for deviations. This is, for cases of this type, the meaning of the thesis that the principal factors determine the investigated phenomena "in the last resort".

It is now evident what mistake is being committed by those who interpret historical materialism as a single-factor theory. They do not understand what the meaning of "determination in the last resort" is (in the

Marxist theory). They treat Marxism as if it assumed – following the positivist principle – that the differentiation of that which is principal and that which is secondary has merely pragmatic importance, showing the investigator's interests. With such an understanding defining as "essential" that which "has many interesting consequences", one may easily see Engels' words as a metaphor. How much ink was wasted in complaints about the metaphorical language of the founders of Marxism (eg., the concept of the last resort)! But the expression "the economic factor determines in the last resort" has a metaphoric meaning within the positivist conceptual framework of the critics alone, not within the theoretical system of the makers of Marxism. It means just that the economic factor is the principal one for social phenomena, and that in its pure form it is accounted for in the basic laws of Marxism, while modified in the numerous concretizations of these laws by multiple minor factors. It is in the same meaning that value determines in the last resort the price of commodities, and the force of inertia – movement of the bodies. Both value and inertia are principal factors deviated from and modified by numerous secondary factors. This view prevents us from considering the founders of Marxism as lacking consistency. As if they were really more radical in some places and in an unanimous and exaggerated way did state that social phenomena depend on the economic factor, whereas in some other places – seeing the difficulties of their own position – retreated to more moderate opinions. As if they expressed either the false, but original view or a more adequate but also not very new one. An introduction to the *A Contribution to the Critique of Political Economy* is quoted as an example of the first case, while the above letter of Engels to Starkenburg is often quoted as an example of the moderate position. It seems, however, that the theorems Marx included in his famous fifteen-sentence long exposition of historical materialism are the basic laws of the Marxist social theory in the meaning presented above. They point out the relation between various forms of social phenomena and their main factors. That is why no word is spoken there about the multifacet dependences of all these phenomena, of the secondary influence of superstructure upon the base, etc. These are all minor effects, which play their role in the applications but not in the exposition of the principles, thus they appear in applications, developments and supplements, for instance in *The Eighteenth Brumaire of Louis Bonaparte* or in letters, but not in the introduction to *A Contribution to the Critique of Political Economy*. There is nothing strange about it. In the abbreviated lecture on classical mechanics one speaks of three principles of the movement of the bodies alone, whereas the cases of movements more complex than the ones which occur when no external forces work are explained by far-fetched concretizations of these principles.

3. BASE AND SUPERSTRUCTURE

Let us consider the fundamental thesis from Marx's *Introduction* saying that base determines superstructure. I shall not bother to define base and superstructure here. My polemics against the critics of Marx should be performed from the methodological standpoint alone, so no discussion will occur here of what interpretation of some concepts in Marxism should be. This thesis is commonly criticised as follows:

Marxian economism – stressing the fact that economic situation is the principal base of any development – is false and cannot be maintained. I think social experience clearly shows that in some conditions the influence of ideas (maybe backed up by propaganda) may overcome and grow stronger than economic forces" (Popper (3), p. 332).

But the thesis we are concerned with may be interpreted as an idealizational statement:

According to the materialist conception of history, the *ultimately* determining factor in history is the production and reproduction of real life. Neither Marx nor I have ever asserted more than this. Hence if somebody twists this into saying that the economic factor is the *only* determining one, he transforms that proposition into a meaningless, abstract, absurd phrase. The economic situation is the basis, but the various elements of the superstructure – political forms of the class struggle and its results . . . , juridicial forms, and especially the reflections of all these real struggles in the brains of the participants, political, legal, philosophical theories, religion views and their further development into systems of dogmas – also exercise their influence upon the course of the historical struggles and in many cases determine their *form* in particular. There is an interaction of all these elements in which, amid all the endless host of accidents (that is, of things and events whose inner interconnection is so remote or so impossible of proof that we can regard it as non-existent and neglect it), the economic movement is finally bound to assert itself (Engels (7), pp. 394–395).

Therefore the thesis that base determines superstructure eliminates all the secondary influences as the reversible influence of superstructure on economic base or the tradition remaining from the previous formations, etc. It is only an approximation of this thesis which can be applied to the actual complicated situations when those secondary influences just occur:

. . . the more distant an investigated economic domain is, and the closer it is to the purely abstract-ideological one, the more accidents are met in its development, the more zigzagged will be the curve of its development. But if an average axis of this curve is drawn it will easily be noted that the longer the period under consideration and the bigger the range of the domain in question, the more its line of development will approach the line of the economic development (Engels (6), p. 483).

As can be seen the arguments of the critics falsify the following thesis: base determines univocally superstructure, while the founders of Marxism stated quite another thesis of an idealizational nature:

(N) if the superstructure does not act on the economic base and no residual of the base and superstructure from previous formations occur, etc., then the economic base of a given formation determines univocally the superstructure of this formation.

This statement does not undergo critical arguments quoted above, since they are concerned with empirical situations when both the principal (economic) and secondary factors occur. At most the arguments may be directed against the approximation of (N):

(N′) the weaker the superstructure act on the economic base and the less there is of the remnants of the base and superstructure from the previous formations, etc., the more precisely does the base of a given formation determine the superstructure in the long run.

A closer inspection would certainly reveal that (N′) is a statistical statement in nature and (N) is semi-idealizational, not simply idealizational (Chapter 11). (N′) dealt with some tendency occurring with greater or lesser intensity depending on numerous interfering factors. It is then theorem which does allow for exceptions (when interfering factors do occur with particular intensity). One cannot falsify it by pointing out the exceptions – which is Popper's argument. If one wants to contradict this statement one should demonstrate that the tendency (and not univocal dependence which is spoken about in the *idealizational law* (N)) does not occur in the history of mankind.

CHAPTER 17

THE LAW OF ABSOLUTE PAUPERIZATION

1. THE OBJECT OF THE CONTROVERSY

It is a very well-known criticism of the Marxian theory which claims that the author of *Capital* "predicted the growing poverty of the masses, whereas the masses in every capitalist country are better off now, than they were a couple of decades ago" (Bober (1), p. 93). This criticism is directed against the so-called law of absolute pauperization which states:

> The greater the social wealth, the functioning capital . . . and, therefore, also the absolute mass of the proletariat and the productiveness of its labour, the greater is the industrial reserve army. . . . But the greater this reserve army in proportion to the active labour-army, the greater is the mass of consolidated surplus-population, whose misery is in direct proportion to the labour put it. The more extensive . . . the industrial reserve army, the greater is official pauperism. *This is the absolute general law of capitalist accumulation* (Marx (1), vol. 1, p. 644).

This statement is usually interpreted as expressing Marx's view that the living standards of the working class will continually go down.

This is also how many Marxists understand it. They invent various arguments against the cirticism in question. Some of them remind Engels' critical remarks on the *Erfurt Program* concerning the growing property of the working class and conclude that at least Engels did not accept the law of absolute pauperization. Some did the same to Lenin's remarks. Some tried to demonstrate that if the capitalist system as a whole is investigated, including the backward countries, then the law of absolute pauperization holds. One had also defended this law by including the intensivity of work within the concept of real wages: the calculations showed that real wages understood in this way drop, though popularly understood real wages in the developed capitalist countries rise. All these defences are based on the assumption that anti-Marxist arguments are correct as far as methodology goes and it is only their contents that needs refutation. But it is precisely their methodology which merits refutation. This lifts the necessity of a content discussion.

2. THE METHODOLOGICAL NATURE OF THE LAW OF ABSOLUTE PAUPERIZATION

The law of absolute pauperization has been introduced – according to the basic Marxian methodological principle – as an idealizational theorem, when some side circumstances were ignored, i.e. – some simplifying assumptions were added, and they allowed us to abstract from these circumstances. In the first volume of *Capital* where the above passage of Marx has been found there is an assumption which was already considered in Chapter two – that the economy in question is a closed one, that it is isolated from foreign contacts. The role of this simplifying assumption is evident. If we assume that the economy is clossed, then we abstract from the possibility of lifting the living standards of the working class by means of colonial exploitation. The model of capitalist society considered in the first volume of *Capital* is composed of two basic classes – industrial capitalists and the workers. And the meaning of this assumption has been explained by Marx himself in the following way. In the first parts of *Capital*:

> The real conditions within which the actual process of production takes place are therefore not analysed. It is assumed throughout, that the commodity is sold at its value. We do not examine the competition of capitals, nor the credit system, nor the actual composition of society, which by no means consists only two classes, workers and industrial capitalists (Marx (5), Part II, pp. 492–493).

The law we are speaking of holds under the assumption that the workers are deprived of class consciousness and do not lead an organized economic struggle against capitalists. But in reality workers, by means such as trade unions, try to organize a planned coordinated action of the employed and unemployed "in order to destroy or to weaken the ruinous effects of this natural law of capitalist production on their class" (Marx (1), Vol. I, p. 640). But, although Marx clearly states that "Every combination of employed and unemployed disturbs the 'harmonious' action of this law" (*ibid.*, p. 640), in the above-quoted formulation of the law of absolute pauperization there is no accounting for the influence of organized trade unionism on wages. Marx had evidently abstracted from this factor as from a secondary one which should be taken into account no sooner than during the concretization of the law.

One could point out many other idealizing assumptions, but the ones that has been mentioned are enough to decide that the law of absolute pauperization is an idealizational statement, as many other Marx's laws are. Therefore the law of absolute pauperization should be formulated as follows:

(1) if the economic system is a closed one and if only two classes exist there: industrial capitalists and the workers, and, moreover, the workers do not lead an organized economic struggle against the capitalists, etc., then the bigger the global capital engaged in this system, the lower real wages of the working class will be.

It is evident at once that the thesis (1) is not contradicted by the argument which the critics of Marxism put forward:

(2) since the end of the 19th century there was an increase of the global capital in the developed capitalist countries, but there was no falldown, to the contrary – an increase in the real wages of the working classes.

The thesis (2) speaks about the empirical capitalist economy, and the law of absolute pauperization (1) speaks of the idealized model of such an economy. The statement (2) cannot, therefore, contradict the law (1) for purely logical reasons. The arguments of the critics of Marxism are based on a misunderstanding, on an improper recognition of the methodological character of the law of absolute pauperization.

3. THE LIMITATIONS OF COUNTER-ARGUMENTATION

The question, however, remains: why has Marx formulated the law of absolute pauperization? The answer – on the basis of the Marxian model of science – is evident: in order to concretize it, to lift the simplifications by accounting for the influence of side circumstances not mentioned in the very formula of the law. The final proof in Marxian writings is that the above formulation of the law of absolute pauperization is accompanied by a sentence: "Like all other laws it is modified in its working by many circumstances, the analysis of which does not concern us here" (Marx (1), Vol. 1, p. 644). In order to 'realize' this law (i.e. to refer it directly to empirical reality) one should introduce, says Marx, the corrections connected with various "varied circumstances" which, to a first approximation were omitted (due to the accepted simplifying assumptions). This is a clear indication that the law in question shall be concretized. It is true, though, that Marx did not concretize it in further fragments of *Capital*, which happened to the law of value, the schemes of simple reproduction or the laws of the rate of surplus value. One may hold it against him, in so far as one may hold anything against anyone who was prevented by death from finishing his work.

But it is quite certain that one cannot attack the Marxian law of pauperization as contrary to empirical reality, which is what most of the critiques of *Capital* do. From the methodological point of view it is as senseless as "falsifying" the inertia principle ("if no force acts upon the body, then it either rests or moves with the uniform and constant movement") just because there happen to exist real bodies which neither rest nor move with a uniform and constant movement, but, let us say, go round in circles.

This answer to the criticism of the Marxian law is the answer of a methodologist and not an economist. The latter should go further than that – account for those minor factors which interfered with the law of pauperization even in Marx's times (colonial exploitation, trade union movement resistance, etc.) and the new ones which Marx could not foresee. No one can guarantee that the procedure of concretization of the law of absolute pauperization will lead to the true theorem. But only when it is completed could one pose a question of the law of pauperization being true or false. The very same question that was considered solved in a negative way for so long.

CHAPTER 18

THE CONTRADICTION BETWEEN THE THIRD AND THE FIRST VOLUME OF *CAPITAL*

1. BÖHM-BAWERK'S CRITICISM

Now I shall discuss another well-known criticism of the Marxian theory. As the economists say, Böhm-Bawerk attack against the Marxian theory of value

> is characteristic and typical for bourgeois economists as far as the Marxist theory is concerned. Whenever a bourgeois economist attempts to criticise Marx's theory he has almost always patterned his arguments along the lines of Böhm-Bawerk, whose criticism had become a classical scheme already (Zawadzki (1), p. 376).

E. Böhm-Bawerk accused Marx of a contradiction between two parts of the theory of value. In the first volume of *Capital* Marx had established the law of value saying that the prices of commodities equal their values. But, according to Böhm-Bawerk, in the third volume of *Capital* he maintains that prices equal not values but prices of production which generally differ from values.

> It is not hard to notice – states the Austrian economist – that the theses and allowances of the third volume clearly contradict the basic doctrines of the first volume [of *Capital* – L.N.]. In the first volume the reader is presented with a logical necessity which develops from the essence of exchange and which makes two different commodities opposing each other in exchange, namely with the fact that they possess only one common feature, that is, labour. In the third volume he learns that the commodities which oppose each other in exchange, in fact, regularly contain, and they have to, different amounts of labour (Böhm-Bawerk (1), p. 90).

Therefore, and this is Böhm-Bawerk's conclusion, "never before had the end of the system contradicted the beginning of it more sharply and briefly" (*ibid.*, p. 90).

2. THE 'CONTRADICTION' AND THE LEVELS OF ABSTRACTION

Let us examine, in the light of my reconstruction of Marx's theory of value (see Chapter 1), two theorems which are supposed to be contradictory. They are statements T^8 and T^5 of Chapter 1, Section 1, that is, the law of value:

(1) if $C(x)$ and $PC(x)$ and $E(x) - I(x) = 0$ and $m'(x) - M'(x) = 0$ and $ORG_A(x) - ORG_I(x) = 0$ and $mer(x) = 0$ and $ORG(x) - ORG_E(x) = 0$ and $D(x) - S(x) = 0$ and $Org(x) - ORG(x) = 0$, then $P(x) = l(V(x))$

and the law of production prices:

(2) if $C(x)$ and $PC(x)$ and $E(x) - I(x) = 0$ and $m'(x) - M'(x) = 0$ and $ORG_A(x) - ORG_I(x) = 0$ and $mer(x) = 0$ and $ORG(x) - ORG_E(x) \neq 0$ and $D(x) - S(x) \neq 0$ and $Org(x) - ORG(x) \neq 0$, then $P(x) = k(x) + k(x)p'(x)$.

Now one may easily observe that the statements do not contradict each other since they speak of different objects. The first one deals with an economy which, among others, contains an ideally balanced market (i.e., demand for each commodity equates supply of it), while the second one is concerned with an economy which, among others, contains a 'realistic market' (i.e., it is allowed that supply is not equal to demand). In the same way two conditionals of different (excluding) antecedents are not contradictory (for instance, the following statements are not contradictory: 'if anyone is Polish, then he or she is Slavic', and 'if anyone is German, then he or she is not Slavic').

In the case under consideration, the statements being accused as being contradictory are idealizational statements of different levels of abstraction since the law of production prices (2) is a concretization of the law of value (1). Therefore their antecedents exclude each other and the conditionals (1) and (2) cannot be accused of being contradictory. They tell us different things but under different conditions.

It is therefore evident that the contradiction which is pointed out by Böhm-Bawerk is impossible on purely logical grounds.

3. HOW THIS CRITICISM WAS POSSIBLE?

The matter seems to be so simple that the question arises how could Böhm-Bawerk and numerous famous economists (e.g., J. Robinson) and philosophers (e.g., K.R. Popper) conclude that the law of value is incompatible with the law of production prices. The explanation also seems to be very simple. They do not understand that the statements in question are idealizational conditionals since they do not interpret Marx's theory as an idealizational theory at all. Hence they treat the theorems in question as factual statements. Under this assumption the contradiction is, in fact, evident. For the statement "prices of commodities are equal to their values" is contradictory to the statement "prices of commodities are equal to their production prices".

It is therefore evident that the Böhm-Bawerk's argumentation is invincible if the theory of value is understood as a factual theory. But the shadow of invincibility disappears when one realizes what the Marxian priciple of theory construction is, i.e., when one realizes that the law of value (1) is an idealizational law and the law of production prices (2) is its concretization. The criticism under consideration was, then, possible only on the condition that the method of idealization occurring in *Capital* had been ignored.

CHAPTER 19

MARX'S HISTORICISM

1. THE CHARGE OF 'HISTORICISM'

'Historicism' has been characterized by K. R. Popper as follows: historicism is a view of social sciences which assumes that their principal aim is a social prognosis, and that this aim can be achieved by analysing the 'rhythms' or 'patterns', 'laws' or 'trends' which the evolution of history undergoes (Popper (2), p. 107). So conceived historicism is represented first of all by Marxism. The basic argument against historicism is that the aim which it puts before itself is impossible to achieve with the means at its disposal. Marx's theorems are not the laws of the type of those formulated in natural sciences since they are descriptions of specific, individual trends. They have, then, the same status as the historical sentence 'Darwin and Galton had a grandfather in common'. This is why because the evolution of life or of human societies is the single, historical process which develops itself according to causal laws but its description is not a law: it is a singular historical sentence (*ibid.*, p. 108). Since it is impossible to explain or predict the facts on the basis of singular statements, the conclusion is that no historical prognosis can be made on the basis of the sentences about historical trends. The illusion, Popper continues, that one can predict on the basis of Marx's theorems is created by his failure of understanding of the nature of scientific predictions (e.g., those made by astronomy, physics, etc.). Marx's historicism arose out of the mixing of scientific prediction and historical prophecy on a great scale. In fact, the type of predictions Marxism presents is closer to the Old Testament with respect to its logical nature than to predictions of modern physics (Popper (4), p. 276).

2. THE ANALYSIS OF MARX'S PREDICTIONS

In order to test whether these arguments are correct let us examine the concrete example of Marx's prediction concerning the situation of the working class after introduction of free commerce. Marx presented this prediction in the discussion with the adherents of free commerce who argued that custom barriers should be abolished in order to improve the situation of the proletariat.

The problem of an influence of complete freedom of commerce upon the situations of the working class is very easy to solve. . . . All the laws referring to this question refer to reality provided that the commerce is free of any bounds and not only in one country but all over the world. These laws . . . become more realistic, precise, they lose their abstract character as free commerce wins. The scientists concerned with these questions always stress that their reasonings are based on the assumption that commerce is being free of any present limitations. And they are quite right assuming this. For it is not making arbitrary abstractions but excluding from the analysis secondary circumstances. Then it may be truly said that economists . . . know more about the society which will be than about the one that is. They know more about the future than about the present (Marx (12), pp. 335–336).

As far as we are concerned with "the law according to which the minimal wage is a natural price of the commodity called labour, this law will be satisfied to the extent to which Ricardo's prediction about free commerce becomes true" (*ibid.*, p. 337). Therefore it is precisely in free commerce that there will appear poverty among the working class, as wages will be equal to the existential minimum.

The question of the truthfulness of this prediction is unimportant from the methodological point of view. I am concerned only with the methodological nature of Marx's predictions since it has been criticized by Popper.

Marx's prediction runs as follows. He starts with the law of minimal wage which he treats as an idealizational statement:

(1) if commercial relations between a given country and all the others countries are entirely free (are subjected to the market rules alone), then the wages in this country are equal to the minimal ones.

Then he proceeds to state that as the custom barriers are lifted which makes it possible to apply the statement to empirical reality through approximation obtaining the following factual statement:

(2) the closer commercial relations of a given country with the other ones are to the ideal of completely free commercial relations, the closer the wages of a given country are to the minimal ones.

And out of statement (1) and of an additional premise

(3) commerce will grow more and more free

Marx draws the prediction:

(4) working class wages will approach the minimal ones.

Notice that Marx, contrary to Popper's thesis, does use law (1) to predict future events. Statement (1) is not a registration of a single historical trend but it is an idealizational statement. Marx's predictions are referring to idealizational laws such as the law of value, the schemes of reproduction (see Chapters 1 and 2), etc. If the idealizational vision of science is assumed they may be said to be "the laws of the type of those of natural sciences." Hence, Marx's predictions are of the same methodological type as those of natural sciences – both resort to idealizational statements and their concretizations or approximations. Methodologically they seem to be of equal worth. Obviously, they are less certain than astronomical or physical predictions which is due to the substantial under-development of social sciences in comparison with natural sciences.

CHAPTER 20

THE CONTRADICTIONS AND AMBIGUITIES WITHIN THE THEORY OF SOCIAL CLASS

1. THE CRITICISM

It is a very popular view that the Marxian theory of social class is rather discouraging with respect to its methodological properties. First of all attention is drawn to Marx's inconsistency: though the theory in question was simple, it was so inadequate that he was forced to change his own dichotomous scheme of a capitalist society to use it in special researchers:

> Marx's model stripped away all status remnants, defining the position of men within capitalist society solely in terms of their relation to the means of production, to the sources of their income . . . . Property as an objective criterion is indispensable to the understanding of the stratification of capitalist society. Alone, it is inadequate and misleading (Mills (1), p. 107).

And further on

> In this [Marx's – L.N.] few comments on the conception of class, as in his general expectations, he stuck to the simplification. In his historical studies, he was more adequate, but on the whole, his simplification become misleading and unfruitful (*ibid.* p. 107).

This gap between the theory of class and its sociological or historical applications is often stressed. Thus, S. Hook writes:

> The term *class* in Marx is used in various senses, not all strictly derivative from each other. Sometimes, its defining feature is the role a group plays in production, sometimes it is their common mode of life, including culture and traditions, sometimes the source of their income or the level of their income, sometimes their vocation or, in the case of of unemployed, their lack of any (Hook (1), p. 39).

Also S. Ossowski states that "One has only to compare the various passages in which the concept of social class is used by either writer [Marx and Engels – L. N.] to realize that the term *class* has for them a variable denotation" (Ossowski (1), p. 71). Ossowski explains this fact in terms of different social roles of the author of *Capital*:

> In his character of revolutionary, economist and sociologist, Marx inherited all three

basic types of conceiving the class structure which are encountered in the history of European thought. These are the dichotomic scheme, the scheme of gradation and the functional scheme (*ibid.*, p. 82).

And so,

Marx the revolutionary and Marx the dramatist of history developed a dichotomic conception of a class society. Marx the sociologist was compelled in his analysis of comtemporary societies to infringe the sharpness of the dichotomic division by introducing intermediate classes (*ibid.*, p. 75).

And finally: "with Marx the theorist, we sometimes have to deal . . . with a scheme which is inherited from bourgeois economics. This is the trichotomous functional scheme of Adam Smith" (*ibid.*, p. 79). If we consider also historical studies of Marx and Engels, "we obtain an image in which the capitalist society is functionally differentiated into seven, eight, or even nine classes or strata" (*ibid.*, p. 82).

2. THE METHODOLOGICAL NATURE OF THE MARXIAN THEORY OF CLASS

The theory of class may be interpreted, as it seems to me, in the same way as the theory of value (see Chapter 1, Section 1) or the theory of reproduction (see Chapter 2, Section 2) or the theory of surplus value (Nowak (2)) or the theory of the rate of profit (Nowak (10)), etc. The theory of class is composed of a sequence of models being in force under decreasing amounts of idealizing assumptions. Each model describes the way a capitalist society is divided and what are the causes and results of this division. In each of these models we deal with a different concept of social class.

Let us discuss some models of the theory of social class to prove the soundness of the above-mentioned criticisms of this theory. Our analysis will be based on *Capital*, the only work of the founders of Marxism where the theory under consideration is presented in a more or less systematic way. More detailed analysis may be found in Jasińska-Nowak (1) and in Nowak (14).

Let us start with noticing that the key concept of the theory of social class, that of a capitalist society, is an idealizational notion in nature. The capitalist mode of production, wrote Marx, "rests on the fact that the material conditions of production are in the hands of non-workers in the form of property and land, while the masses are only owners of the personal condition of production of labour power" (Marx (9), p. 25). But this definition presents the capitalist society in its 'pure form': ' "Present-day society' is a capitalist society which exists in all civilized countries, more or less free from medieval admixture, more or less modified by the special historical development of each country, more or less developed" (*ibid.*, p. 32). This notion of a class society is characterized with the aid of some idealizing assumptions. Let us try to reconstruct some of them (more about this see Nowak (14) and (15)).

The idealizing condition of a great importance is expressed in *Capital* in the following way: ". . . we treat the capitalist producer as owner of the entire surplus-value" (Marx (1), Vol. I, p. 565). The assumption is necessary because "the simple fundamental form of the process of accumulation is obscured . . . by the splitting up of surplus-value. An exact analysis of the process, therefore, demands that we should, for a time, disregard all phenomena that hide the play of inner mechanism" (*ibid.*, p. 565). The point is that an industrial capitalist in fact shapes a part of the surplus value with other types of capitalists such as merchants (who put his product into circulation), bankers (who give him loans and credits) or landowners (who own the land which is used for production). But at the beginning Marx disregards all this assuming that the entire surplus value is equal to the profit of industrial capitalists. This means that according to Marx's assumption, all other "forms of surplus value" as commercial profit, interest and ground rent equal zero. The sociological content of this condition is, then, that there are no merchants, bankers or landowners in a capitalist society in the idealized sense.

Some other idealizing conditions are adopted only tacitly. It is the theory contained in *Capital* which is dealt with the mature capitalist mode of production. But the latter, according to many of Marx's explanations, presupposes that production is conducted on a large scale. This means that small-scale capitalists are not taken into account. Also within the theory of *Capital*, Marx presupposes that wages of workers are more or less equal – he admits their differentiation only in historical sections of *Capital* which are used to approximate some of idealizational statements of his theory. Similarly, Marx does not acknowledge differentiation of members of social categories he is concerned with from the viewpoint of modes of life or cultural patterns. In other words, he tacitly assumes that all the members of a particular category are similar in these respects.

Another tacit assumption in *Capital* is that the social categories in question lack class consciousness and organization, and that they have no 'literary' or 'political' representatives, i.e. people who conduct ideological or political activity in their interest. The classes under consideration are regarded initially as 'classes in themselves'.

Another idealizing assumption is that the capitalist society which is studied in *Capital* is free of social classes which linger on from previous socio-economical formations:

In order to examine the object of our consideration in its integrity, free from all disturbing subsidiary circumstances, we must . . . assume that capitalist production is everywhere established and has possesed itself of every branch of industry '(Marx (1), Vol. I, p. 581).

Let us list some of the assumptions of the Marxian theory of class (more about this see Nowak (14) and Jasińska-Nowak (1)):

p_0: *S* is a capitalist society,

p_1: society *S* does not apply any pre-capitalist forms of production,

p_2: all social categories in society *A* lack 'political representatives',

p_3: all social categories in society *S* lack 'ideological representatives',

p_4: all people in society *S* are on a similar cultural level and follow a similar mode of life within a particular social category of this society,

p_5: industrial producers in society *S* (among others, capitalist farmers) do not pay any ground rent for their land (i.e., they own it themselves),

p_6: in society *S*, industrial producers do not borrow money (i.e., they use their own capital only),

p_7: in society *S*, industrial producers market their products themselves.

3. THE MARXIAN THEORY OF SOCIAL CLASS: AN ATTEMPT AT A RECONSTRUCTION

All the assumptions adopted in Marx's theory of class are far from being listed in the above enumeration. The latter has to be enforced with additional conditions as the assumption of all social categories being mutually exclusive. Nonetheless the conditions presented above should be sufficient for my purposes in this brief presentation of the idealizational interpretation of the Marxian theory of social class.

(a) Model I

Here is Marx's formulation of his basic law of class division of a capitalist society:

Apart from this class [industrial capitalists – L.N.] there is, according to our assumption – the general and exclusive domination of capitalist production – no other class but the working class (Marx (1), Vol. II, p. 401).

This 'assumption' confirms that the law in question may be interpreted as an idealizational one. In the initial, basic model of the theory of class there are two classes only: industrial capitalists (including the ones who are engaged in agriculture since Marx used the term 'industrial' in its economic, not colloquial, meaning) and industrial workers.

Let us consider the criterion of class division which is applied in the highest level of abstraction. As Marx explains in the first volume of *Capital*, ". . . our 'proletarian' is economically none other than the wage-labourer who produces and increases capital" (Marx (1), Vol. I, p. 614). As it follows from this, an industrial capitalist is someone whose capital the worker increases. This being so, the following definition of a social class has to be adopted in Model I:

(Df.I) class_1 is a group of people who either produce surplus value, but do not own it, or do not produce surplus value, but own it.

From the definition of a social class, the law of the social structure of a capitalist society has to be distinguished:

(I.1) if p_0 and. . . and p_7, then society S is divided into industrial capitalists and industrial workers.

Industrial capitalists are, under economic theory of *Capital*, people who do not produce but appropriate surplus value, while industrial workers are the ones who produce surplus value but do not receive it.

Apart from the problem of class division of the capitalist society, the question of its genesis arises within the theory of social class. And Marx's answer is that the appropriation of surplus value is determined by social ownership of the means of production with which workers produce this surplus value:

Wherever a part of society possesses the monopoly of the means of production, the labourer, free or not free, must add to the working-time in order to produce the means of subsistence for the owners of the means of production" (*ibid.*, p. 235).

This idea may be reconstructed thus:

(I.2) if p_0 and. . . and p_7, then owners of the means of production in society *S* appropriate surplus value, while the owners of the labour power who are deprived of the means of production receive value indispensable for the reconstruction of their labour power.

And also the theorem about results of class division belongs to the Marxian theory: since "profit rises to the extent that wages fall, it falls to the extent that wages rise", then "the interests of capital and the interests of wage labour are diametrically opposed" (Marx (4), pp. 96–97). Hence the following idealizational statement holds in Model I:

(I.3) if p_0 and ... and p_7, then the larger part of surplus value goes to industrial capitalists, the less do industrial workers get, and the smaller the part appropriated by industrial capitalists, the higher do wages rise.

This statement may be termed the law of contradiction of objective interests of industrial capitalists and workers.

Certainly, there occur also other statements within Model I about relationships between industrial capitalists and workers in terms of power and in terms of consciousness, etc., (see Jasińska–Nowak (1)), but we shall concentrate here on only these matters which seem to be important for rejection of criticisms of the theory of class mentioned at the beginning of this chapter.

(b) Model II

In Model II, assumption p_7 is removed. Starting from Chapter XVI of the third volume of *Capital*, Marx examines a society in which industrial capitalists no longer market their own products, but there is a specialised group of people (called merchants) who do it. As Marx says,

> in the course of scientific analysis, the formation of a general rate of profit appears to result from industrial capitals and their competition, and is only later corrected, supplemented and modified by the intervention of merchant's capital (Marx (1), Vol. III, p. 287).

This "intervention" consists in the following:

> The function of selling it [a commodity – L.N.] . . . has passed from the manufacturer to the merchant, has become special business of the merchant, whereas previously it was a function which the producer had to perform himself after having completed the function of its production (*ibid.*, p. 270).

Obviously, such words as 'after' and 'previously' do not signify any chronological order in this context, but refer to the order of concretization.

In this connect on Marx introduces a new category of commercial capitalists as well as of the commercial workers they employ. The latter are employed in the sphere of circulation and hence they do not produce surplus value. Nevertheless, "the unpaid labour of these clerks, while it does not create surplus-value, enables [a commercial capitalist – L.N.] to appropriate surplus-value . . . . It is therefore a source of profit for him" (*ibid.*, p. 294). So industrial and commercial capitalists are the owners of profit produced for them by workers (industrial and commercial correspondingly). This leads to the change of definition of a capitalist from the first volume of *Capital*:

The dealer in commodities, as a capitalist generally, appears on the market primarily as a representative of a certain sum of money„ which he advances as a capitalist, i.e. which he wants to turn from x (its original value) into $x + \Delta x$ (the original sum plus profit) (*ibid.*, p. 296).

Hence, the new definition of a class is valid in Model II:

(Df.II) class$_2$ is a group of people who receive profit without producing it with their own labour, or do not receive it although they produce it with their labour.

So conceived classes are the following:

(II.1) if p_0 and . . . and p_7, then society S is divided into 'active capitalists' and workers.

'Active capitalists' are those who receive profit (not simply surplus value as in Model I) without producing it, while workers are those who do not receive the profit they produce with their labour. So defined social categories include groups of people highly heterogeneous sociologically, but in view of accepted idealizing assumptions such matters are not taken into account within Model II. Therefore 'active capitalists' include both industrial capitalists and merchants while workers include both manual workers and skilled salesmen in specialized bookshops.

When commercial capitalists come into the picture, industrial capitalists receive only a part of surplus value. This part is called an industrial profit, while the remaining part is termed a commercial profit. Hence theorem (I. 2) is modified within Model II:

(II.2) if p_0 and . . . and p_6 and non-p_7, then owners of the means of production in society S appropriate industrial profit, owners of the means of circulation appropriate commercial profit and labourers employed by both receive wages.

The modification of the theorem (I.3) may be in turn, presented thus:

(II.3) if p_0 and ... and p_6 and non-p_7, then (a) if the commercial profit is constant, then the higher the profit of industrial capitalists in society *S* rises, the more the wages of industrial labourers fall, and vice versa, (b) if the profit of industrial capitalists is constant, then the higher the profit of commercial capitalists in *S* rises, the lower the wages of commercial labourers fall, and vice versa, (c) in society *S* "the larger the merchant's capital in proportion to the industrial capital, the smaller the rate of industrial profit, and vice versa" (*ibid.*, p. 266).

This theorem provides the basis not only for conflict between industrial capitalists and industrial workers (as (I.3) does) but also between commercial capitalists and commercial workers and industrial capitalists and commercial ones. Hence the image of social conflicts in Model II is richer than in Model I only because it provides the basic conflict of a capitalist society.

(c) Model III

In part V of the third volume of *Capital* Marx removes assumption p_6. He introduces the concept of interest whose sociological meaning is that the new category of people becomes an object of his consideration: money capitalists who lend money and derive their profit from interest. Besides them Marx introduces also the category of bankers who concentrate "large amounts of the loanable money-capital... so that, in place of the individual money-lender, the bankers confront the industrial capitalists and commercial capitalists as representatives of all money-lenders" (*ibid.*, p. 402). Bankers belong to the category of "active capitalists" in a new extended sense, while money capitalists are not 'active capitalists' in this sense. Hence the concept of the class is modified now:

(Df.III) $class_3$ is a group of people who appropriate profit without producing it themselves, or receive interest on capital, or neither receive the profit they produce nor take interest on loaned capital.

Banker's employees are referred to as banking workers and their status is exactly the same as that of commercial workers – they do not produce surplus value but do provide bankers with profit. So we shall use the term 'worker' in an extended sense to include also banking workers. Therefore the statement about class division is modified as follows in Model III:

(III.1) if p_0 and ... and p_5 and non-p_6 and non-p_7, then society S is divided into capitalists and workers.

This statement is a modification of (II.1), because the category of capitalists (people who appropriate profit or interest which they do not produce with their own labour) includes the category of 'active capitalists' in the sense of Model II plus the category of bankers (the former and the latter make the category of 'active capitalists' in the sense of Model III) and plus the category of money capitalists. Similarly, workers in the present sense includes workers in the sense of Model II and banking workers.

The origin of the part of profit which goes to money capitalists is the following: industrial capitalists find that it pays to operate with somebody else's capital for which they pay interest to its owners (either directly or through a bank). So interest is a part of surplus value that industrial capitalists transfer to other capitalists whose sole occupation is to loan capital (according to our assumption it is maintained that this category excludes each other, e.g. the category of 'active capitalists' which in reality must not be true). On the other hand, bankers derive their profit from the discount rate. This being the case, economic foundations of class division may be presented as follows:

(III.2) if p_0 and ... and p_5 and non-p_6 and non-p_7, then the owners of the means of production in society S appropriate industrial profit, the owners of the means of circulation – commercial profit, the owners of loaned capital – interest, bankers derive their profit from the discount rate, while labourers employed by these categories of capitalists receive wages.

Let us consider now the mutual relationships between the distinguished social categories. Money capitalists and workers do not have opposing interests: a growth of interest cuts surplus value but does not affect workers' wages. Hence, there is (an objective) opposition of interests between all the categories distinguished in Model II and also between money capitalists and 'active capitalists' and between industrialists and merchants on the one hand, and the bankers, on the other. Let us express all this in the following statement:

(III.3) if p_0 and ... and p_5 and non-p_6 and non-p_7, then (a) [the same as in (II.3)(A)], (b) [the same as in (II.3)(b)], (c) [the same as in (II.3)(c)], (d) if the discount rate is constant, then the higher the profit of bankers rises, the lower the wages of banking workers fall, and vice versa (e) "assuming the average profit to be given, the rate of the profit of enterprise is not determined by wages, but by the rate of interest. It is high or low in inverse proportion to it" (*ibid.*, p. 379).

(d) Model IV

Next Marx removes, at the end of his manuscript, assumption p_5 which leads to the introduction of a new social category, namely landowners. In this connection the criterion of class division is modified. Landowners cut back the profits of a sub-category of capitalists, namely capitalist farmers, who have to pay the landowners for the right to invest capitals into the land the latter own. Now one may observe that within Model IV the criterion of class division is the ownership of a source of income:

> The owners merely of labour-power, owners of capital, and land-owners, whose respective sources of income are wages, profit and ground rent, in other words wage-labourers, capitalists and landowners, constitute then three big classes of modern society based upon the capitalist mode of production (*ibid.*, p. 885).

Hence, within Model IV the following notion of class is valid:

(Df.IV) class_4 is a group of people who own the same source of income.

Obviously, the notion of a source of income should be understood here in the sense of the Marxian political economy, not in the colloquial sense. Therefore, wage, profit and ground rent are different sources of income, while a physician and a clerk may have the same source of income.

The statement concerning the class division is modified in Model IV in the following way:

(IV.1) if p_0 and ... and p_4 and non-p_5 and non-p_6 and non-p_7, then society S is divided into capitalists, landowners and labourers,

while the economic foundation of this division presents itself in Model IV as follows:

(IV.2) if p_0 and ... and p_4 and non-p_5 and non-p_6 and non-p_7, then in society S owners of the means of production appropriate industrial profit, owners of the means of circulation – commercial profit, owners of loanable capital – interest, landowners – ground rent, while labourers receive wages.

Marx's manuscript of *Capital* breaks off after the part in which he formulates these statements concerning the class division and its causes. So Model IV is the last stage of concretization which can be found in *Capital*. Likely, many others corrections and supplements would have been introduced to proceed this simple image of the class structure near to 19th century societies. Some of these corrections have been tentatively reconstructed in Jasińska–Nowak (1) and in Nowak (14).

(e) Further Models

Let us try to reconstruct in a brief form further notions of class starting from $class_4$ in order to recognize the notion of class employed in Marx's and Engels' political and historical writings.

One may remove assumption p_4 taking into account some sociological characteristics of people. In this way one may proceed to Marx's definition of class from *The Eighteenth Brumaire of Louis Bonaparte* which mentions the following properties of class (in our terms – $class_5$): "mode of life", "interests" and "culture" (Marx (10), p. 334). These features are not accidentally pointed out but they are supposed to be determined by economic conditions of people, that is, above all by the nature of a source of income:

> Insofar as millions of families live under conditions of existence which separate their mode of life, their interests and their culture from those of other classes, and put them in hostile opposition to the latter, they form a class (*ibid.*, p. 334).

So, within the next model one would introduce the following concept of a class:

(Df.V) $class_5$ is a group of people with the same source and similar level of income, closely related economic interests, mode of life and culture.

Notice that Marx and Engels usually used the terms 'bourgeoisie' and 'proletariat' as names of particular $classes_5$. That must have been a conscious choice, if it is considered that in the 1891 edition of Marx's study *Wage, Labour and Capital* (first published in 1849), Engels consistently replaced the word 'bourgeois' with 'capitalist' wherever there was talk exclusively about economic relationships between the enterpreneur and the worker: while he left unchanged the passages where Marx spoke of the sociological characteristics of the capitalist class. This would imply that Marx and Engels used the word 'bourgeoisie' as a more realistic name of the class, reserving the more abstract notion of capitalist for the theoretical analyses. An analysis of Marx's and Engels' writings in which the term 'bourgeois' appears shows that they use it to denote big and medium capitalists such as industrial, commercial, money, bankers and landowners (Marx (11), pp. 139–140).

After removing p_3 the following notion of a class may be introduced:

(Df.VI) class$_6$ is formed of those members of class$_5$ who are class-conscious,

while after removing p_2 the distinction between a class-in-itself and class-for-itself becomes possible. The definition of class within postulated Model VII:

(Df.VII) class$_7$ is class$_6$ which has political representation

would correspond to class-for-itself, while class$_6$ without political representation (and perhaps even class$_5$ without class consciousness) would correspond to a class-in-itself.

All these notion of class are in force under condition p_1, i.e., that the capitalist system free of residual elements of previous social classes is considered. By removing this assumption one may introduce the category of peasants, as well as the category of craftsmen. Both two categories correspond to many passages in Marx's and Engels' historical and political works (see Jasińska–Nowak (1) and Nowak (14)).

(f) An Analysis of the Criticism

If the presented interpretation, which is a very brief ouline only, is correct (see more detailed studies already quoted) then the criticisms we have started with may be overthrown. On the assumption that the theory of class is an idealizational one, the charge of inconsistency falls flat. Since the dichotomous grasp of the class structure is expressed in the statement:

(I.1) if p_0 and ... and p_7, then society S is divided into industrial capitalists and industrial workers

while in their historical works Marx and Engels employed another statement, for instance:

(VIII.1) if p_0 and non-p_1 and ... and non-p_7, then society S is divided into the bourgeosie, petty bourgeoisie, peasants and the proletariat.

These statements are not contradictory at all. The reason is the same as in the case of the law of value and the law of production prices – they are statements of different levels of abstraction.

Also another charge saying that Marx used the term 'class' in quite different meanings seems to be soundless. If we take a closer look at the concepts of class appearing in model I to VIII, that is, concepts class_1 to class_8, we shall see that they include all the concepts mentioned by S. Hook or S. Ossowski. However, these notions are interconnected with the specific relation which is termed in the language of the founders of Marxism a developments of a notion. One concept is a development of another one, if it is less abstract (i.e. it has been introduced with a lesser amount of idealizing conditions) and its content is a modification of the concent of the previous one with respect to the removed conditions. Each of such concepts is introduced for a given model and is valid only within it. The chain of notions is then established corresponding to the chain of models constituting a given theory. This is what one may observe in the case of sequence of notions class_1 to class_8: these concepts are linked with the relation of development.

If this methodological principle of introducing notions is understood (see also Chapter 4, Section 1(b)), then one cannot criticise Marx for applying the term 'class' in different meanings. To the contrary: one could accuse him of inconsistency if he applied the same concept of class in subsequent models, i.e., at different levels of abstraction. The modification of the notion is necessary if one proceeds from the idealized model to a more realistic one.

It should be added that Marx criticised other authors for not developing concepts they applied, that is, for being inconsistent in his own sense. This is what he wrote on the epoch-making work of Ricardo:

Thus one can see that in this first chapter not only are *commodities* assumed to exist – and when considering value as such, nothing further is required – but also wages, capital, profit, the general rate of profit and even . . . the various forms of capital as they arise from the process of circulation, and also the difference between 'natural and market-price' (Marx (5), Vol. II, p. 168).

And this criticism of Ricardo is further generalized by him as follows:

[Ricardo] consciously *abstracts* from the form of competition, from the appearance of competition, in order to comprehend the *laws as such*. . . . one must criticise him for regarding the phenomenal form as immediate and direct proof or exposition of the general laws, and for failing to *interpret* it (*ibid.*, p. 106).

One cannot therefore stop at definition of such a concept (category) which is suitable for a given model, but giving up gradual simplifications one should develop this concept.

The conclusion of this chapter is similar to the previous ones: the criticism against the conceptual framework of the Marxian class theory results from the interpretation of this theory from the standpoint of one's own methodological principles, and not from the point of view of the Marxian methodology according to which this theory had been constructed. Or, more properly, not from the point of view of the idealizational interpretation of the Marxian methodology presented in this book.

REFERENCES CITED

Ajdukiewicz, K.: (1) *Pragmatic Logic*, Reidel, Dordrecht, 1974.

Ajdukiewicz, K.: (2) *Language and Logic* (in Polish), Vol. 1, Warsw, 1960.

Allen, R.G.M.: (1) *Mathematical Analysis for Economists*, Macmillan, London, 1949.

Allen, R.G.M.: (2) *Mathematical Economy*, Macmillan, London, 1959.

Aristotle, *Physics* (Polish translation), Warsaw, 1968.

Banaszak, G.: (1) 'Classification in the Idealizational Concept of Science' (in Polish), in: L. Nowak (Ed)., 'Idealization and Adaptation', *Poznańskie Studia z Filozofii Nauki, 4*, Warsaw-Poznan, 1979.

Bober, M.H.: (1) *Karl Marx's Interpretation of History*, Harvard University Press, Cambridge, 1950.

Böhm-Bawerk, E.: (1) *Capital and the Profit of Capital* (Polish translation), Vol. II, Gebethner and Wolff, Cracow, 1924.

Bridgeman, P.W.: (1) *The Logic of Modern Physics*, Macmillan, London, 1948.

Brzeziński, J.: (1) 'Empirical Essentialist Procedures in Behavioral Enquiries', in: Idealizational Concept of Science, *Poznań Studies in the Philosophy of the Sciences and the Humanities*, Vol. 2, No. 3, special issue, 1976.

Brzeziński, J.: (2) *The Structure of Behavioral Science* (in Polish), Warsaw-Poznań, 1976.

Brzeziński, J.: (3) 'Essential Structure, Interaction, Experiment,' *Poznań Studies in the Philosophy of the Sciences and the Humanities*, Vol. 1, No. 1, 1975.

Brzeziński, J.: (4) 'The Structure of Proto-idealizational Theories' (in Polish), in: L. Nowak (Ed)., 'Theory and Reality' (in Polish), *Poznańskie Studia z Filozofii Nauki, 1*, Warsaw-Poznań, 1976.

Brzeziński, J., Burbelka, I., Klawiter, A., Magala, S., and Nowak, L.: (1) 'Law and Theory. A Contribution to an Idealizational Interpretation of the Marxist Methodlogy', in: 'Idealizational Concept of Science', *Poznań Studies in the Philosophy of the Sciences and the Humanities*, Vol. 2, No. 3, 1976.

Bunge, M.: (1) *Causality. The Place of the Causal Principle in Modern Science*, The World Publishing Company, Cleveland—New York, 1963.

Carnap, R.: (1) *The Logical Structure of the World*, Routledge and Kegan Paul, London, 1968.

Carnap, R.: (2) *Philosophy and Logical Syntax*, Routledge and Kegan Paul, London, 1935.

Carnap, R.: (3) 'Foundations of Logic and Mathematics', *International Encyclopedia of Unified Science*, Vol. I, No. 3, 1955, 3rd edn.

Carnap, R.: (4) *The Methodological Character of Theoretical Concepts*, (*Minnesota Studies in the Philosophy of Science*), Vol. I, Minneapolis, 1956.

Carnap, R., Hahn, H., and Neurath, 0.: (1) 'Scientific Outlook' (Polish translation), in: H. Buczyńska (Ed)., Vienna Circle, Warsaw, 1960.

Chwalisz, P.: (1) 'Constants within the Idealizational Theory of Science, in: A. Klawiter and L. Nowak (Eds.). *Truth, Evidence, Discovery, History, and Idealization* (in Polish), Warsaw-Poznań, 1979.

Chwalisz, P., Kowalik, P., Nowak, L., Patryas, W., and Stefański, M.: (1) 'Peculiarities of Practical Research, in: 'Idealizational Concept of Science', *Poznań Studies in the*

Philosophy of the Sciences and the Humanities , Vol. 2, No. 3, special issue, 1976.
Czerwiński, Z.: (1) *Introduction to Econometrics* (in Polish), Poznań, 1967.
Einstein, A.: (1) *The Essence of Relativity* (Polish translation), Warsaw, 1962.
Engels, F.: (1) 'Marx's Capital, in: F. Engels, *On Marx's Capital*, Progress Publishers, Moscow, 1967.
Engels, F.: (2) 'Letter to Schmidt' (March 12, 1895), in: K. Marx and F. Engels, *Selected Correspondence*, Progress Publishers, Moscow, 1971.
Engels, F.: (3) Dialektik der Natur, Dietz Verlag, Berlin, 1951.
Engels, F.: (4) 'Letter to Lavrov' (November 12, 1875), in: K. Marx and F. Engels, *Selected Correspondence*, Progress Publishers, Moscow, 1971.
Engels, F.: (5) Ludwig Feuerbach and the Eclipse of the German Classical Philosophy, in: K. Marx and F. Engels, *Selected Works*, Vol. III, Progress Publishers, Moscow, 1970.
Engels, F.: (6) 'Letter to Starkenburg' (January 25, 1894), in: K. Marx and F. Engels, *Selected Works* (Polish translation), Vol. II, Warsaw, 1949.
Engels, F.: (7) 'Letter to Bloch' (September 21, 1890), in: K. Marx and F. Engels, *Selected Correspondence*, Progress Publishers, Moscow, 1971.
Feyerabend, P.K.: (1) *Explanation, Reduction, and Empiricism*, (*Minnesota Studies in the Philosophy of Science*), Vol. III, Minneapolis, 1957.
Frish, S. and Timoreva, A.: *Course in Physics*, Vol. 1, (Polish translation), Warsaw, 1962.
Galileo, G.: (1) *Dialogue of Two World Systems* (Polish translation), Warsaw, 1962.
Galileo, G.: (2) *Mathematical Considerations* (Polish translation), Warsaw, 1931.
Gibson, Q.: (1) *The Logic of Social Enquiry*, Routledge and Kegan Paul, London, 1960.
Hempel, C. G.: (1) 'Problems of Concepts and Theory Formation in the Social Sciences,' in: *Science, Language and Human Rights*, Philadelphia, 1952.
Hempel, C. G.: (2) 'The Theoretician's Dilemma,' in: D. Shapere (Ed.),. *Philosopkical Problems of Natural Science*, Collier-Macmillan, New York, London, 1965.
Hempel, C. G.: (3) *Philosophy of Natural Sciences*, Englewood Cliffs, N. J., 1968.
Hook, S.: (1) *Marx and the Marxists*, Van Nostrand Co., New York, 1955.
Jasińska, A., and Nowak, L.: (1) 'Foundations of Marx's Theory of Class: A Reconstruction,' *Poznań Studies in the Philosophy of the Sciences and the Humanities*, Vol. 1, No. 1, 1975.
Kalecki, M.: (1) 'Notes on the Theory of Growth,' in: *Theories of Economic Growth and the Development of Contemporary Capitalism* (in Polish), Warsaw, 1962.
Kitajgorodski, A.: (1) *Physics* (Polish translation), Warsaw, 1965.
Klawiter, A.: (1) 'Kuhn and Feyerabend on Scientific Facts and Their Interpretation,' *Poznań Studies in the Philosophy of the Sciences and the Humanities*, Vol. 1, No. 2, 1975.
Klein, L. R. : (1) *Introduction to Econometrics*, M.I.T. Press, Cambridge Mass., 1961.
Kline, R.: (1) *Mathematics and the Physical World* (Polish translation), Warsaw, 1961.
Kmita, J.: (1) 'Remarks on Marxian Holism,' in: *Methodological Assumptions of Marx's Capital* (in Polish), Warsaw, 1970.
Kmita, J.: (2) *Methodological Problems of the Humanistic Interpretation* (in Polish), Warsaw 1971.
Kmita, J., and Nowak, L.: (1) *Studies in the Theoretical Foundations of the Humanities* (in Polish), Poznań, 1968.

Kmita, J., and Nowak, L.: (2) 'The Assumption of Rationality in Human Sciences,' *The Polish Sociological Bulletin*, 1970, No. 1.

Kocikowski, A.: (1) 'Idealization and Testing' in: L. Nowak, (Ed.), 'Idealization and Adaptation' (in Polish), *Poznańskie Studia z Filozofii Nauki*, 4, Warsaw-Poznań, 1977.

Korsch, K.: (1) *Marxism and Philosophy*, NLB, London, 1970.

Kotarbińska, J.: (1) 'The Controversy: Inductivism – Deductivism,' in: T. Pawłowski (Ed.), *The Logical Theory of Science* (in Polish), Warsaw, 1967.

Kozówna, D.: (1) 'Pedagogics as a Practical Discipline', in: L. Nowak (Ed.), 'Theory and Reality' (in Polish), *Poznańskie Studia z Filozoffi Nauki* 4, 1977.

Krajewski, W.: (1) 'On Notions of Relative Truth', *Studia Filozoficzne*, 1963, Nos. 3–4

Kuhn, T. S.: (1) *The Structure of Scientific Revolutions, International Encyclopaedia of Unified Science*, Chicago, 1962.

Kuhn, T.S.: (2) *The Copernican Revolution. Planetary Astronomy in the Development of Western Thought*, Harvard University Press, Harvard 1957.

Lakatos, I.: (1) 'Falsificationism and the Methodology of Scientific Research Programmes', in: I. Lakatos and A. Musgrave (Eds)., *Criticism and the Growth of Knowledge*, Cambridge University Press, Cambridge, 1970.

Lange, O.: (1) *Political Economy* (in Polish), Vol. 1, Warsaw, 1967.

Lange, O.: (2) 'A New Version of Kalecki's Model', in: O. Lange, *Social and Economic Works* (in Polish), Warsaw, 1960.

Lenin, W.I.: (1) *Philosophical Notebooks* (Polish translation), Warsaw, 1956.

Lukács, G.: (1) *History and Class Consciousness*, M.I.T. Press, Cambridge, Mass., 1968.

Marx, K,: (1) *Capital*, Vols. I–III, Progress Publishers, Moscow, 1968.

Marx, K.: (2) Letter to Engels (April 2, 1858), in: K. Marx, F. Engels, Selected Correspondence, Progress Publishers, Moscow, 1971.

Marx, K.: (3) 'Preface to "A Contribution to a Critique of Political Economy"', in: K. Marx and F. Engels, *Selected Works*, Vol. I, Progress Publishers, Moscow, 1962.

Marx, K.: (4) 'Wages, *Price and Profit*', in: *ibid.* Vol. II.

Marx, K.: (5) *Theories of Surplus Value*, Vols. I–III, Progress Publishers, Moscow, 1968.

Marx, K.: (6) 'Letter to Kugelmann' (July 11, 1868), in: K. Marx and F. Engels, *Selected Correspondence*, Progress Publishers, Moscow, 1971.

Marx, K.: (7) *Grundrisse. Foundations of the Critique of Political Economy*, Penguin Books, London, 1973.

Marx, K.: (8) 'Letter to Engels' (March 25, 1868), in: K. Marx and F. Engels, *Selected Correspondence*, Progress Publishers, Moscow, 1971.

Marx, K.: (9) 'Critique of the Gotha Programme', in: K. Marx and F. Engels, *Selected Works*, Vol. I, Progress Publishers, Moscow, 1962.

Marx, K.: (10) 'The Eighteenth Brumaire of Louis Bonaparte', in: *ibid.*

Marx, K.: (11) 'The Class Struggle in France', in: *ibid.*

Marx, K.: (12) 'On Free Trade', in: K. Marx and F. Engels, *Collected Works* (in Polish), Vol. 4, Warsaw, 1962.

Massalski, J., and Studnicki, J.: (1) *International System of Units of Measurement SI* (in Polish), Warsaw, 1971.

Mills, C.W.: (1) *The Marxists*, Laurel, New York, 1962.

Morawski, S.: (1) 'Criteria of Evaluation of a Work of Art' (in Polish), *Studia Estetyczne*, Vol. IV, 1964.

Naganowski, T.: (1) personal communication.

Nagel, E.: (1) *The Structure of Science*, Routledge and Kegan Paul, London, 1961.

Nowakowa, I.: (1) 'Idealization and the Problem of Correspondence', *Poznań Studies in the Philosophy of the Sciences and the Humanities* Vol. 1, No. 1, 1975.

Nowakowa, I.: (2) 'Partial Truth, Relative Truth, Absolute Truth', in 'Categorial Interpretation of Dialectics', *Poznań Studies in the Philosophy of the Sciences and the Humanities*, Vol. 2, No. 4, special issue, 1976.

Nowakowa, I.: (3) *Dialectical Correspondence and the Growth of Science* (in Polish), Warsaw-Poznań, 1975.

Nowakowa, I.: (4) 'On a Concept of Correspondence', *Poznań Studies in the Philosophy of the Sciences and the Humanities*, Vol. 1, No. 2, 1975.

Nowakowa, I.: (5) *Problems of the Theory of Truth in the Marxist Philosophy* (in Polish), Poznań 1977

Nowak, L.: (1) 'On the Principle of Abstraction and Gradual Concretization', in: *Methodological Assumptions of Marx's 'Capital'* (in Polish), Warsaw, 1970.

Nowak, L.: (2) *Foundations of the Marxian Methodology* (in Polish), Warsaw, 1971.

Nowak, L.: (3) *Principles of the Marxist Philosophy of Science. An Attempt at a Systematic Reconstruction* (in Polish), Warsaw, 1974.

Nowak, L.: (4) The Problem of Explanation in Marx's 'Capital', *Quality and Quantity*, Vol. V, No. 1, 1971.

Nowak, L.: (5) 'Idealization: A Reconstruction of Marx's Ideas', *Poznań Studies in the Philosophy of the Sciences and the Humanities*, Vol. 1, No. 1, 1975.

Nowak, L.: (6) 'Relative Truth, the Correspondence Principle, Absolute Truth', *Philosophy of Science*, Vol. 42, No. 2, 1975.

Nowak, L.: (7) 'On Some Interpretation of Marxist Methodology', *Zeitschrift für allgemeine Wissenschaftstheorie*, Vol. VII/1, 1976.

Nowak, L.: (8) *Lectures of the Marxist Philosophy*, Vol. I: *Categorial Interpretation of Dialectics*, Poznań, 1976, Vol. II: *Categorial Ontology and Idealizational Epistemology*, Poznań, 1977 (both in Polish).

Nowak, L.: (9) 'The Problem of Adequacy of Idealization in the Idealizational Concept of Science' (in Polish, in print).

Nowak, L.: (10) 'Laws of Science, Theory, Measurement', *Philosophy of Science*, Vol. 39, No. 4, 1972.

Nowak L.: (11) On the Nature of the Marxist Dialectics', *Dialectics and Humanism*, Vol. I, No. 1, 1974.

Nowak, L.: (12) 'Evaluation and Cognition', in: 'Value, Valuation, and Science', *Poznań Studies in the Philosophy of the Sciences and the Humanities*, Vol. 2, No. 1, special issue, 1976.

Nowak, L.: (13) 'Value, Idealization, and Valuation,' *Quality and Quantity*, Vol. 8, 1974, 107–120.

Nowak, L.: (14) 'Marx's Model of the Structure of Capitalist Society (in Publish), *Studia Socjologiczne*, 1972, No. 2.

Nowak, L.: (15) 'The Anatomy of Anti-Marxist Criticism,' *Poznań Studies in the Philosophy of the Sciences and the Humanities*, Vol. 1, No. 3, 1975.

Ossowski, S.: (1) *Class Structure in Social Consciousness*, Routledge and Kegan Paul, London, 1963.

Pap, A.: (1) *An Introduction to the Philosophy of Science*, Routledge and Kegan Paul, 1963.

Patryas, W.: (1) 'An Analysis of the "Ceteris Paribus" Clause', *Poznań Studies in the Philosophy of the Sciences and the Humanities*, Vol. 1, No. 1, 1975.

Patryas, W.: (2) 'Idealization and Empirical Testing', in: *Poznań Studies in the Philosophy of the Sciences and the Humanities*, Vol. 3, 1977.

Patryas, W.: (3) *Experiment and Idealization* (in Polish), Warsaw-Poznań, 1976.

Pawlowski, Z.: (1) *Introduction to Econometrics* (in Polish), Warsaw, 1964.

Pawlowski, Z.: (2) *Theory of Econometric Prediction* (in Polish), Warsaw, 1969.

Piekara, A.: (1) *General Mechanics* (in Polish), Warsaw, 1964.

Popper, K. R.: (1) *The Logic of Scientific Discovery*, Hutchinson, London, 1959.

Popper, K. R. (2) *The Poverty of Historicism*, Routledge and Kegan Paul, London, 1967.

Popper, K. R. (3) *Conjectures and Refutations*, Routledge and Kegan Paul, London 1965.

Popper, K. R. (4) 'Prediction and Prophecy in the Social Sciences', in: P. Gardiner, *Theories of History*, Macmillan, London, 1959.

Przełecki, M.: (1) 'Operationism', *Studia Logica*, Vol. III.

Rescher, N.: (1) *Topics in Philosophical Logic*, Reidel, Dordrecht, 1968.

Robinson, J.: (1) *Studies on the Marxian Economics* (Polish translation), Warsaw, 1961.

Rozental, M.: (1) Dialektika "Kapitata" Karta Marksa (in Russian), Moscow 1967.

Rudner, R.: (1) *Philosophy of Social Sciences*, Englewood Cliffs, N. J., 1966.

Samuelson, P.: (1) *Economics* (Russian translation), Moscow, 1965.

Schlick, M.: (1) *Gesammelte Aufsätze*, Vienna, 1938.

Sorokin, P.: (1) *Social Theories of the 19th and 20th* centuries (Polish translation), Cracow (undated).

Stefański, M.: (1) 'Similarity, factor, vagueness', in: L. Nowak (Ed.), Assumptions of Dialectics' (in Polish), *Poznańskie Studia z Filozofii Nauki* 2, Warsaw-Poznań, 1977.

Such, J.: (1) 'Idealization and Concretization in Natural Sciences, *Poznań Studies in the Philosophy of the Sciences and the Humanities*, (in print).

Suszko, R.: (1) 'Syntactic Structures and Semantic Reference,' *Studia Logica*, Vol. VII–VIII.

Szefler, M.: (1) Political Economy of Socialism (in Polish), Warsaw, 1970.

Tuchańska, B.: (1) 'A Phenomenon, the Essence, a Theory,' *Poznań Studies in the Philosophy of the Sciences and the Humanities*, Vol. 2, No. 1, 1976.

Tuchańska, B.: (2) 'Factor vs. Magnitude,' in: 'Idealizational Concept of Science,' *Poznań Studies in the Philosophy of the Sciences and the Humanities*, Vol. 2, No. 3, special issue, 1976.

Tuchańska, B.: (3) 'Idealization and Measurement', *Poznań Studies in the Philosophy of the Sciences and the Humanities*, Vol. 3, 1977.

Tuchańska, B.: (4) 'Factor, Magnitude, Dependence. An Attempt at Concretization of the Idealizational Concept of Science', (in Polish, in print).

Watkins, J. W. N.: (1) 'Ideal Types and Historical Explanation', in: H. Feigl and M. Boodbeck (Eds.), *Readings in the Philosophy of Science*, Appleton-Century-Crofts, Inc., New York, 1953.

Weber, M.: (1) *The Methodology of the Social Sciences*, The Free Press, Glencoe Ill., 1949.

von Wright, G.H.: (1) *The Logic of Preference*, Edinburgh, 1963.

Wójcicki, S.: (1) *The Principles of Experimentation in Engineering* (in Polish), Warsaw, 1963.

Zawadzki, J.: (1) *Political Economy of Capitalism*, Vol. 1, (in Polish), Warsaw, 1967.

Zielińska, R.: (1) 'Analogous of the Idealization in Mathematics,' in: L. Nowak (Ed.),' Theory and Reality' (in Polish), *Poznańskie Studia z Filozofii Nauki* 1, Warsaw-Poznań, 1976.

Zielińska R.: (2) 'On Inter-functional Concretization', in: 'Idealizational Concept of Science', *Poznań Studies in the Philosophy of the Sciences and the Humanities*, Vol. 2, No. 3, special issue 1976.

Znamierowski, Cz.: (1) 'Ethics, Normative Again,' in: *Charisteria* (in Polish), Warsaw, 1960.

NAME INDEX

SUBJECT INDEX

SYNTHESE LIBRARY

Studies in Epistemology, Logic, Methodology, and Philosophy of Science

1. J. M. Bochénski, *A Precis of Mathematical Logic*. 1959.
2. P. L. Guiraud, *Problèmes et méthodes de la statistique linguistique*. 1960.
3. Hans Freudenthal (ed.), *The Concept and the Role of the Model in Mathematics and Natural and Social Sciences*. 1961.
4. Evert W. Beth, *Formal Methods. An Introduction to Symbolic Logic and the Study of Effective Operations in Arithmetic and Logic*. 1962.
5. B. H. Kazemier and D. Vuysje (eds.), *Logic and Language. Studies Dedicated to Professor Rudolf Carnap on the Occasion of His Seventieth Birthday*. 1962.
6. Marx W. Wartofsky (ed.), *Proceedings of the Boston Colloquium for the Philosophy of Science 1961-1962*. Boston Studies in the Philosophy of Science, Volume I. 1963.
7. A. A. Zinov'ev, *Philosophical Problems of Many-Valued Logic*. 1963.
8. Georges Gurvitch, *The Spectrum of Social Time*. 1964.
9. Paul Lorenzen, *Formal Logic*. 1965.
10. Robert S. Cohen and Marx W. Wartofsky (eds.), *In Honor of Philipp Frank*. Boston Studies in the Philosophy of Science, Volume II. 1965.
11. Evert W. Beth, *Mathematical Thought. An Introduction to the Philosophy of Mathematics*. 1965.
12. Evert W. Beth and Jean Piaget, *Mathematical Epistemology and Psychology*. 1966.
13. Guido Küng, *Ontology and the Logistic Analysis of Language. An Enquiry into the Contemporary Views on Universals*. 1967.
14. Robert S. Cohen and Marx W. Wartofsky (eds.), *Proceedings of the Boston Colloquium for the Philosophy of Science 1964-1966. In Memory of Norwood Russell Hanson*. Boston Studies in the Philosophy of Science, Volume III. 1967.
15. C. D. Broad, *Induction, Probability, and Causation. Selected Papers*. 1968.
16. Günther Patzig, *Aristotle's Theory of the Syllogism. A Logical-Philosophical Study of Book A of the Prior Analytics*. 1968.
17. Nicholas Rescher, *Topics in Philosophical Logic*. 1968.
18. Robert S. Cohen and Marx W. Wartofsky (eds.), *Proceedings of the Boston Colloquium for the Philosophy of Science 1966-1968*. Boston Studies in the Philosophy of Science, Volume IV. 1969.

19. Robert S. Cohen and Marx W. Wartofsky (eds.), *Proceedings of the Boston Colloquium for the Philosophy of Science 1966-1968.* Boston Studies in the Philosophy of Science, Volume V. 1969.
20. J. W. Davis, D. J. Hockney, and W. K. Wilson (eds.), *Philosophical Logic.* 1969.
21. D. Davidson and J. Hintikka (eds.), *Words and Objections. Essays on the Work of W. V. Quine.* 1969.
22. Patrick Suppes, *Studies in the Methodology and Foundations of Science. Selected Papers from 1911 to 1969.* 1969.
23. Jaakko Hintikka, *Models for Modalities. Selected Essays.* 1969.
24. Nicholas Rescher *et al.* (eds.), *Essays in Honor of Carl G. Hempel. A Tribute on the Occasion of His Sixty-Fifth Birthday.* 1969.
25. P. V. Tavanec (ed.), *Problems of the Logic of Scientific Knowledge.* 1969.
26. Marshall Swain (ed.), *Induction, Acceptance, and Rational Belief.* 1970.
27. Robert S. Cohen and Raymond J. Seeger (eds.), *Ernst Mach: Physicist and Philosopher.* Boston Studies in the Philosophy of Science, Volume VI. 1970.
28. Jaakko Hintikka and Patrick Suppes, *Information and Inference.* 1970.
29. Karel Lambert, *Philosophical Problems in Logic. Some Recent Developments.* 1970.
30. Rolf A. Eberle, *Nominalistic Systems.* 1970.
31. Paul Weingartner and Gerhard Zecha (eds.), *Induction, Physics, and Ethics.* 1970.
32. Evert W. Beth, *Aspects of Modern Logic.* 1970.
33. Risto Hilpinen (ed.), *Deontic Logic: Introductory and Systematic Readings.* 1971.
34. Jean-Louis Krivine, *Introduction to Axiomatic Set Theory.* 1971.
35. Joseph D. Sneed, *The Logical Sstructure of Mathematical Physics.* 1971.
36. Carl R. Kordig, *The Justification of Scientific Change.* 1971.
37. Milic Capek, *Bergson and Modern Physics.* Boston Studies in the Philosophy of Science, Volume VII. 1971.
38. Norwood Russell Hanson, *What I Do Not Believe, and Other Essays* (ed. by Stephen Toulmin and Harry Woolf). 1971.
39. Roger C. Buck and Robert S. Cohen (eds.), *PSA 1970. In Memory of Rudolf Carnap.* Boston Studies in the Philosophy of Science, Volume VIII. 1971.
40. Donald Davidson and Gilbert Harman (eds.), *Semantics of Natural Language.* 1972.
41. Yehoshua Bar-Hillel (ed.), *Pragmatics of Natural Languages.* 1971.
42. Sören Stenlund, *Combinators, λ-Terms and Proof Theory.* 1972.
43. Martin Strauss, *Modern Physics and Its Philosophy. Selected Papers in the Logic, History, and Philosophy of Science.* 1972.
44. Mario Bunge, *Method, Model and Matter.* 1973.
45. Mario Bunge, *Philosophy of Physics.* 1973.
46. A. A. Zinov'ev, *Foundations of the Logical Theory of Scientific Knowledge (Complex Logic).* (Revised and enlarged English edition with an appendix by G. A. Smirnov, E. A. Sidorenka, A. M. Fedina, and L. A. Bobrova.) Boston Studies in the Philosophy of Science, Volume IX. 1973.
47. Ladislav Tondl, *Scientific Procedures.* Boston Studies in the Philosophy of Science, Volume X. 1973.
48. Norwood Russell Hanson, *Constellations and Conjectures* (ed. by Willard C. Humphreys, Jr.). 1973.

49. K. J. J. Hintikka, J. M. E. Moravcsik, and P. Suppes (eds.), *Approaches to Natural Language*. 1973.
50. Mario Bunge (ed.), *Exact Philosophy – Problems, Tools, and Goals*. 1973.
51. Radu J. Bogdan and Ilkka Niiniluoto (eds.), *Logic, Language, and Probability*. 1973.
52. Glenn Pearce and Patrick Maynard (eds.), *Conceptual Change*. 1973.
53. Ilkka Niiniluoto and Raimo Tuomela, *Theoretical Concepts and Hypothetico-Inductive Inference*. 1973.
54. Roland Fraïssé, *Course of Mathematical Logic* – Volume 1: *Relation and Logical Formula*. 1973.
55. Adolf Grünbaum, *Philosophical Problems of Space and Time*. (Second, enlarged edition.) Boston Studies in the Philosophy of Science, Volume XII. 1973.
56. Patrick Suppes (ed.), *Space, Time, and Geometry*. 1973.
57. Hans Kelsen, *Essays in Legal and Moral Philosophy* (selected and introduced by Ota Weinberger). 1973.
58. R. J. Seeger and Robert S. Cohen (eds.), *Philosophical Foundations of Science*. Boston Studies in the Philosophy of Science, Volume XI. 1974.
59. Robert S. Cohen and Marx W. Wartofsky (eds.), *Logical and Epistemological Studies in Contemporary Physics*. Boston Studies in the Philosophy of Science, Volume XIII. 1973.
60. Robert S. Cohen and Marx W. Wartofsky (eds.), *Methodological and Historical Essays in the Natural and Social Sciences. Proceedings of the Boston Colloquium for the Philosophy of Science 1969-1972*. Boston Studies in the Philosophy of Science, Volume XIV. 1974.
61. Robert S. Cohen, J. J. Stachel, and Marx W. Wartofsky (eds.), *For Dirk Struik. Scientific, Historical and Political Essays in Honor of Dirk J. Struik*. Boston Studies in the Philosophy of Science, Volume XV. 1974.
62. Kazimierz Ajdukiewicz, *Pragmatic Logic* (transl. from the Polish by Olgierd Wojtasiewicz). 1974.
63. Sören Stenlund (ed.), *Logical Theory and Semantic Analysis. Essays Dedicated to Stig Kanger on His Fiftieth Birthday*. 1974.
64. Kenneth F. Schaffner and Robert S. Cohen (eds.), *Proceedings of the 1972 Biennial Meeting, Philosophy of Science Association*. Boston Studies in the Philosophy of Science, Volume XX. 1974.
65. Henry E. Kyburg, Jr., *The Logical Foundations of Statistical Inference*. 1974.
66. Marjorie Grene, *The Understanding of Nature. Essays in the Philosophy of Biology*. Boston Studies in the Philosophy of Science, Volume XXIII. 1974.
67. Jan M. Broekman, *Structuralism: Moscow, Prague, Paris*. 1974.
68. Norman Geschwind, *Selected Papers on Language and the Brain*. Boston Studies in the Philosophy of Science, Volume XVI. 1974.
69. Roland Fraïssé, *Course of Mathematical Logic* – Volume 2: *Model Theory*. 1974.
70. Andrzej Grzegorczyk, *An Outline of Mathematical Logic. Fundamental Results and Notions Explained with All Details*. 1974.
71. Franz von Kutschera, *Philosophy of Language*. 1975.
72. Juha Manninen and Raimo Tuomela (eds.), *Essays on Explanation and Understanding. Studies in the Foundations of Humanities and Social Sciences*. 1976.

73. Jaakko Hintikka (ed.), *Rudolf Carnap, Logical Empiricist. Materials and Perspectives.* 1975.
74. Milic Capek (ed.), *The Concepts of Space and Time. Their Structure and Their Development.* Boston Studies in the Philosophy of Science, Volume XXII. 1976.
75. Jaakko Hintikka and Unto Remes, *The Method of Analysis. Its Geometrical Origin and Its General Significance.* Boston Studies in the Philosophy of Science, Volume XXV. 1974.
76. John Emery Murdoch and Edith Dudley Sylla, *The Cultural Context of Medieval Learning.* Boston Studies in the Philosophy of Science, Volume XXVI. 1975.
77. Stefan Amsterdamski, *Between Experience and Metaphysics. Philosophical Problems of the Evolution of Science.* Boston Studies in the Philosophy of Science, Volume XXXV. 1975.
78. Patrick Suppes (ed.), *Logic and Probability in Quantum Mechanics.* 1976.
79. Hermann von Helmholtz: *Epistemological Writings. The Paul Hertz/Moritz Schlick Centenary Edition of 1921 with Notes and Commentary by the Editors.* (Newly translated by Malcolm F. Lowe. Edited, with an Introduction and Bibliography, by Robert S. Cohen and Yehuda Elkana.) Boston Studies in the Philosophy of Science, Volume XXXVII. 1977.
80. Joseph Agassi, *Science in Flux.* Boston Studies in the Philosophy of Science, Volume XXVIII. 1975.
81. Sandra G. Harding (ed.), *Can Theories Be Refuted? Essays on the Duhem-Quine Thesis.* 1976.
82. Stefan Nowak, *Methodology of Sociological Research. General Problems.* 1977.
83. Jean Piaget, Jean-Blaise Grize, Alina Szeminska, and Vinh Bang, *Epistemology and Psychology of Functions.* 1977.
84. Marjorie Grene and Everett Mendelsohn (eds.), *Topics in the Philosophy of Biology.* Boston Studies in the Philosophy of Science, Volume XXVII. 1976.
85. E. Fischbein, *The Intuitive Sources of Probabilistic Thinking in Children.* 1975.
86. Ernest W. Adams, *The Logic of Conditionals. An Application of Probability to Deductive Logic.* 1975.
87. Marian Przelecki and Ryszard Wójcicki (eds.), *Twenty-Five Years of Logical Methodology in Poland.* 1977.
88. J. Topolski, *The Methodology of History.* 1976.
89. A. Kasher (ed.), *Language in Focus: Foundations, Methods and Systems. Essays Dedicated to Yehoshua Bar-Hillel.* Boston Studies in the Philosophy of Science, Volume XLIII. 1976.
90. Jaakko Hintikka, *The Intentions of Intentionality and Other New Models for Modalities.* 1975.
91. Wolfgang Stegmüller, *Collected Papers on Epistemology, Philosophy of Science and History of Philosophy.* 2 Volumes. 1977.
92. Dov M. Gabbay, *Investigations in Modal and Tense Logics with Applications to Problems in Philosophy and Linguistics.* 1976.
93. Radu J. Bogdan, *Local Induction.* 1976.
94. Stefan Nowak, *Understanding and Prediction. Essays in the Methodology of Social and Behavioral Theories.* 1976.
95. Peter Mittelstaedt, *Philosophical Problems of Modern Physics.* Boston Studies in the Philosophy of Science, Volume XVIII. 1976.

96. Gerald Holton and William Blanpied (eds.), *Science and Its Public: The Changing Relationship*. Boston Studies in the Philosophy of Science, Volume XXXIII. 1976.
97. Myles Brand and Douglas Walton (eds.), *Action Theory*. 1976.
98. Paul Gochet, *Outline of a Nominalist Theory of Proposition. An Essay in the Theory of Meaning*. 1980. (Forthcoming.)
99. R. S. Cohen, P. K. Feyerabend, and M. W. Wartofsky (eds.), *Essays in Memory of Imre Lakatos*. Boston Studies in the Philosophy of Science, Volume XXXIX. 1976.
100. R. S. Cohen and J. J. Stachel (eds.), *Selected Papers of Léon Rosenfeld*. Boston Studies in the Philosophy of Science, Volume XXI. 1978.
101. R. S. Cohen, C. A. Hooker, A. C. Michalos, and J. W. van Evra (eds.), *PSA 1974: Proceedings of the 1974 Biennial Meeting of the Philosophy of Science Association*. Boston Studies in the Philosophy of Science, Volume XXXII. 1976.
102. Yehuda Fried and Joseph Agassi, *Paranoia: A Study in Diagnosis*. Boston Studies in the Philosophy of Science, Volume L. 1976.
103. Marian Przelecki, Klemens Szaniawski, and Ryszard Wójcicki (eds.), *Formal Methods in the Methodology of Empirical Sciences*. 1976.
104. John M. Vickers, *Belief and Probability*. 1976.
105. Kurt H. Wolff, *Surrender and Catch: Experience and Inquiry Today*. Boston Studies in the Philosophy of Science, Volume LI. 1976.
106. Karel Kosík, *Dialectics of the Concrete*. Boston Studies in the Philosophy of Science, Volume LII. 1976.
107. Nelson Goodman, *The Structure of Appearance*. (Third edition.) Boston Studies in the Philosophy of Science, Volume LIII. 1977.
108. Jerzy Giedymin (ed.), *Kazimierz Ajdukiewicz: The Scientific World-Perspective and Other Essays, 1931-1963*. 1978.
109. Robert L. Causey, *Unity of Science*. 1977.
110. Richard E. Grandy, *Advanced Logic for Applications*. 1977.
111. Robert P. McArthur, *Tense Logic*. 1976.
112. Lars Lindahl, *Position and Change. A Study in Law and Logic*. 1977.
113. Raimo Tuomela, *Dispositions*. 1978.
114. Herbert A. Simon, *Models of Discovery and Other Topics in the Methods of Science*. Boston Studies in the Philosophy of Science, Volume LIV. 1977.
115. Roger D. Rosenkrantz, *Inference, Method and Decision*. 1977.
116. Raimo Tuomela, *Human Action and Its Explanation. A Study on the Philosophical Foundations of Psychology*. 1977.
117. Morris Lazerowitz, *The Language of Philosophy. Freud and Wittgenstein*. Boston Studies in the Philosophy of Science, Volume LV. 1977.
118. Stanislaw Leśniewski, *Collected Works* (ed. by S. J. Surma, J. T. J. Srzednicki, and D. I. Barnett, with an annotated bibliography by V. Frederick Rickey). 1980. (Forthcoming.)
119. Jerzy Pelc, *Semiotics in Poland, 1894-1969*. 1978.
120. Ingmar Pörn, *Action Theory and Social Science. Some Formal Models*. 1977.
121. Joseph Margolis, *Persons and Minds. The Prospects of Nonreductive Materialism*. Boston Studies in the Philosophy of Science, Volume LVII. 1977.
122. Jaakko Hintikka, Ilkka Niiniluoto, and Esa Saarinen (eds.), *Essays on Mathematical and Philosophical Logic*. 1978.
123. Theo A. F. Kuipers, *Studies in Inductive Probability and Rational Expectation*. 1978.

124. Esa Saarinen, Risto Hilpinen, Ilkka Niiniluoto, and Merrill Provence Hintikka (eds.), *Essays in Honour of Jaakko Hintikka on the Occasion of His Fiftieth Birthday*. 1978.

125 Gerard Radnitzky and Gunnar Andersson (eds.), *Progress and Rationality in Science*. Boston Studies in the Philosophy of Science, Volume LVIII. 1978.

126. Peter Mittelstaedt, *Quantum Logic*. 1978.

127. Kenneth A. Bowen, *Model Theory for Modal Logic. Kripke Models for Modal Predicate Calculi*. 1978.

128. Howard Alexander Bursen, *Dismantling the Memory Machine. A Philosophical Investigation of Machine Theories of Memory*. 1978.

129. Marx W. Wartofsky, *Models: Representation and the Scientific Understanding*. Boston Studies in the Philosophy of Science, Volume XLVIII. 1979.

130. Don Ihde, *Technics and Praxis. A Philosophy of Technology*. Boston Studies in the Philosophy of Science, Volume XXIV. 1978.

131. Jerzy J. Wiatr (ed.), *Polish Essays in the Methodology of the Social Sciences*. Boston Studies in the Philosophy of Science, Volume XXIX. 1979.

132. Wesley C. Salmon (ed.), *Hans Reichenbach: Logical Empiricist*. 1979.

133. Peter Bieri, Rolf-P. Horstmann, and Lorenz Krüger (eds.), *Transcendental Arguments in Science. Essays in Epistemology*. 1979.

134. Mihailo Marković and Gajo Petrović (eds.), *Praxis. Yugoslav Essays in the Philosophy and Methodology of the Social Sciences*. Boston Studies in the Philosophy of Science, Volume XXXVI. 1979.

135. Ryszard Wójcicki, *Topics in the Formal Methodology of Empirical Sciences*. 1979.

136. Gerard Radnitzky and Gunnar Andersson (eds.), *The Structure and Development of Science*. Boston Studies in the Philosophy of Science, Volume LIX. 1979.

137. Judson Chambers Webb, *Mechanism, Mentalism, and Metamathematics. An Essay on Finitism*. 1980. (Forthcoming.)

138. D. F. Gustafson and B. L. Tapscott (eds.), *Body, Mind, and Method. Essays in Honor of Virgil C. Aldrich*. 1979.

139. Leszek Nowak, *The Structure of Idealization. Towards a Systematic Interpretation of the Marxian Idea of Science*. 1979.

140. Chaim Perelman, *The New Rhetoric and the Humanities. Essays on Rhetoric and Its Applications*. 1979.

141. Wlodzimierz Rabinowicz, *Universalizability. A Study in Morals and Metaphysics*. 1979.

MIX
Papier aus verantwortungsvollen Quellen
Paper from responsible sources
FSC® C105338
FSC
www.fsc.org

If you have any concerns about our products,
you can contact us on
ProductSafety@springernature.com

In case Publisher is established outside the EU,
the EU authorized representative is:
Springer Nature Customer Service Center GmbH
Europaplatz 3, 69115 Heidelberg, Germany

Printed by Libri Plureos GmbH
in Hamburg, Germany